MATLAB App Development

Ganesh R. Naik

MATLAB App Development

A Practical Guide to Building Interactive GUIs

Springer

Ganesh R. Naik
Centre for Artificial Intelligence Research
and Optimisation (AIRO)
Torrens University
Adelaide, SA, Australia

ISBN 978-981-95-8796-4 ISBN 978-981-95-8797-1 (eBook)
https://doi.org/10.1007/978-981-95-8797-1

This Springer imprint is published by the registered company Springer Nature Singapore Pte Ltd.
The registered company address is: 152 Beach Road, #21-01/04 Gateway East, Singapore 189721, Singapore

Preface

Modern engineering, science, and data-driven disciplines increasingly rely on interactive software tools to explore data, test ideas, and communicate results effectively. MATLAB Apps provide a powerful way to transform algorithms, models, and analyses into intuitive, user-friendly applications. By combining computation with carefully designed Graphical User Interfaces (GUIs), MATLAB App Designer enables researchers, educators, and practitioners to create professional-grade Apps without the overhead of traditional software development.

MATLAB App Designer has become the standard environment for building Apps within the MATLAB ecosystem. It offers an integrated workflow for designing interfaces, managing user interactions, visualizing data, and deploying standalone or web-based applications. Whether the goal is rapid prototyping, teaching, research dissemination, or industrial deployment, Apps play a crucial role in making computational tools accessible to a broader audience.

This book is designed as a hands-on guide to designing, developing, and deploying MATLAB Apps using App Designer. It is written for students, researchers, and professionals with a basic familiarity with MATLAB who wish to move beyond scripts and functions toward fully interactive applications. The book covers both foundational and advanced topics, ranging from essential User Interface (UI) components and layout design to event-driven programming, interactive visualizations, performance optimization, and App packaging.

With the growing demand for intuitive analytical tools and decision-support systems, MATLAB Apps are increasingly used in education, engineering design, finance, data analysis, and applied research. Rather than treating Apps as an add-on, this book presents App development as a core skill that bridges the gap between computation and real-world usability.

How This Book is Organized

The book is structured into five parts that progressively build your App development skills. Early chapters introduce the App Designer environment, core UI components, and layout strategies. Subsequent chapters focus on writing App code, handling user input, managing App logic, and ensuring robust behavior. Advanced chapters explore interactive visualizations, performance considerations, and packaging and distributing Apps. The final part presents complete, real-world Apps that demonstrate how the concepts come together in practice.

Each chapter begins with a brief conceptual overview, followed by practical examples developed step by step in MATLAB App Designer. Code snippets and design decisions are explained in context, helping readers understand not just *what* to do, but *why* it is done that way.

Why This Book is Timely

While MATLAB has long been used for numerical computing and algorithm development, there is comparatively little structured material explicitly focused on **modern App Designer-based App development**. Many existing resources rely on legacy GUIDE workflows or focus solely on scripting. App Designer represents a significant shift toward integrated, object-oriented, and event-driven App development within MATLAB.

As interactive tools become increasingly important for teaching, research, and deployment, there is a clear need for a practical, example-driven guide that reflects current best practices. This book aims to fill that gap by providing a clear and structured pathway from basic App design to fully deployable applications.

How to Use This Book

This book is intended to be used actively. Readers are encouraged to recreate the Apps presented in each chapter, modify layouts, experiment with callbacks, and extend the examples to suit their own applications. There are no formal problem sets; instead, learning is driven by exploration, experimentation, and iterative improvement, an approach that closely mirrors real App development workflows.

About the Code/Apps

All examples in this book are developed using recent versions of MATLAB and App Designer. To maintain clarity, non-essential code has been omitted from some listings, keeping attention on the core concepts being discussed. Complete, runnable versions of the Apps are available online (in the Springer GitHub repository), allowing readers to explore, modify, and extend the examples as needed.

Using the App Examples

Readers are encouraged not only to run the Apps, but to adapt them: change layouts, add new UI components, modify callback logic, or integrate your own data and algorithms. The skills developed through customization and experimentation will be far more valuable than simply reproducing the examples.

Resources

All the App examples, supporting files, and updates related to this book are available through the Springer GitHub repository.

Final Note

By the time readers complete reading this book, they will have developed a strong practical foundation in MATLAB App Designer, enabling them to create robust, interactive, and deployable Apps. I hope that this book will serve not only as a guide to *MATLAB App Development* but also as a catalyst for creating tools that make computational work more accessible, engaging, and impactful.

Adelaide, Australia
January 2026

Ganesh R. Naik

Competing Interests The author has no competing interests to declare that are relevant to the content of this manuscript.

Contents

Part I
Introduction

Chapter 1
Introduction

1.1 Introduction

MATLAB is a powerful high-level programming language that engineers and scientists use for developing sophisticated systems. To convey these systems to others, a user-friendly Graphical User Interface (GUI) often represents the most convincing way. App Designer in MATLAB motivates its users to create various Apps by smoothly changing complex ideas into intuitive, visually attractive GUIs [1, 2].

The MATLAB App is an extension of the MATLAB GUI. App Designer is an intuitive development environment; it combines the MATLAB Editor, offering a full palette of interactive User Interface (UI) components [3, 4]. It allows users to design and program the layout of an App and to lay out the UI using a programmable grid layout manager. Automatic reflow options ensure that the App scales to different screen sizes. You can also create the Apps by packaging them into installer files directly from the App Designer Toolstrip. Users can build standalone desktop or web applications (with MATLAB Compiler™) and distribute their App creations with ease.

Apps are self-contained programs in the MATLAB environment. Apps have a user interface for carrying out tasks or computations [5, 6].

Key Features of MATLAB APPs:

- **All-in-One Solution**: Apps encompass all necessary operations within their interface:
  - Data input and acquisition.
  - Data processing and calculations.
  - Result visualization and output.
- **Accessibility**: Many MATLAB products include pre-built Apps.
- **Customization**: You can design and create your own custom Apps using the App Designer development environment.

G. R. Naik, *MATLAB App Development*, https://doi.org/10.1007/978-981-95-8797-1_1

MATLAB Apps provide an efficient and easy way to access MATLAB's computational power. With App Designer, you can create your own custom Apps. These can be used to customize workflows and automate tasks. In this chapter, we introduce the main concepts and advantages of MATLAB Apps. In the following chapters, more information will be provided on how to develop MATLAB Apps using App Designer, more advanced approaches to designing graphical interfaces, and guidelines for producing and distributing Apps [7, 8].

1.2 Why Do We Need MATLAB Apps?

We create MATLAB Apps for a variety of compelling reasons:

- **User-Friendliness**:
  - Simplified Interaction: Apps provide an intuitive graphical interface, making it easier for users (even those with limited coding experience) to interact with complex MATLAB functionality.
  - Guided Workflow: Apps can guide users through a specific set of steps, ensuring consistency and reducing the risk of errors.
- **Efficiency and Productivity**:
  - Automation: Apps automate repetitive tasks, saving significant time and effort.
  - Streamlined Analysis: Apps can be designed to perform complex analyses or simulations with minimal user input.
  - Rapid Prototyping: Apps facilitate rapid prototyping and exploration of different scenarios.
- **Collaboration and Sharing**:
  - Easy Distribution: Apps can be easily packaged and shared with colleagues, collaborators, or even the broader MATLAB community.
  - Reproducibility: Apps promote reproducibility of research or engineering results by capturing the entire workflow in a single, shareable package.
- **Customization**:
  - Tailored Solutions: Apps can be customized to specific needs and workflows, providing a tailored solution for individual users or teams.
  - Domain-Specific Tools: Apps can be created for specific domains, such as signal processing, image analysis, and control systems, providing powerful tools for experts in those fields.
- **Enhanced Visualization**:
  - Interactive Results: Apps can incorporate interactive visualizations, allowing users to explore data and results dynamically.

  - Improved Communication: Visual representations often convey information more effectively than raw data, making it easier to communicate findings to others [5, 7, 8].

By creating MATLAB Apps, we can leverage MATLAB's computational capabilities to deliver a user-friendly, efficient experience for developers and end users alike [7].

1.3 Where to Get MATLAB Apps from?

MATLAB Apps can be obtained in two ways:

- **MATLAB Product Bundles**: Many MATLAB toolboxes, such as Curve Fitting Toolbox™, Signal Processing Toolbox™, and Control System Toolbox™, include pre-built Apps. You can easily access these Apps within the "Apps" gallery, which displays the Apps available with your installed products.
- **Create Your Own Apps (App Designer)**:
  - **App Designer**: The recommended environment for developing custom MATLAB Apps.
  - **Single-File Packaging**: App Designer allows you to package your App into a single file for convenient distribution.
  - **Automated Dependency Handling**: The App packaging tool automatically identifies and includes all necessary files and dependencies, including required MATLAB products.
  - **Sharing Options**:

    Direct Sharing: Distribute your App directly to other users.
    MATLAB File Exchange: Share your App with the broader MATLAB community by uploading it to the File Exchange.

  - **Simplified Installation**: Users can easily install your App without worrying about MATLAB's search path or other complex installation details [7, 9].

1.4 Creating MATLAB Apps—Best Practices

To create high-quality MATLAB Apps, it is essential to following several best practices. These guidelines include the following:

1. **User Interface (UI) Design**:
   - Keep it Simple: Prioritize a clean and intuitive UI. Avoid clutter and unnecessary complexity.
   - Consistent Layout: Ensure consistent use of spaces, fonts, and color schemes in the App for a professional and user-friendly experience.

- Meaningful Labels: Labels on GUI elements (like buttons, text boxes) must be clear and to the point.
- Provide Feedback: You should provide the user with visible indicators, such as progress bars or status messages, to let them know what progress is being made by the App.
- Error Handling: Utilize effective error-handling techniques to handle unforeseen circumstances efficiently, along with providing meaningful error messages to the users.

2. **Code Quality**:

- Modular Design: This involves structuring the App functions in terms of well-defined functions and subroutines.
- Comments and Documentation: Add comments to your code that describe the purpose of differing sections, and add comments to document your App functionality.
- Code Style: Follow proper guidelines regarding coding style (e.g., indentation standards) for better readability and maintainability.
- Testing: It is essential to testing your app thoroughly with different input conditions to detect and remove possible bugs. You could also consider unit testing or integration testing to thoroughly test your App.

3. **Performance**:

- Optimize Code: Optimize code for performance, especially for computationally intensive tasks.
- Efficient Algorithms: Choose efficient algorithms and data structures to minimize processing time.
- Parallel Computing: Utilize parallel computing techniques (e.g., Parallel Computing Toolbox) to accelerate computations whenever possible [7–9].

1.5 Overview of the Book

This book serves as a comprehensive guide to creating interactive, user-friendly applications in MATLAB. Whether you are a beginner or an experienced MATLAB user, this book will equip you with the knowledge and skills to effectively leverage App Designer for a wide range of applications.

Part 1: Introduction sets the stage by introducing the core concepts of MATLAB Apps and their significance. You will learn about the key benefits of using Apps and gain an overview of the App Designer interface.

Part 2: App Designer Fundamentals delves into the foundational aspects of App Designer. You will explore the essential UI components, learn how to design effective layouts, and understand the basic principles of writing App code.

Part 3: App Logic and Behavior focuses on enhancing the functionality of your Apps. You will learn how to handle user input effectively, implement data validation, and create robust error-handling mechanisms to ensure a seamless user experience.

Part 4: Advanced App Development explores advanced techniques for creating sophisticated, interactive Apps. This includes creating custom UI components, developing interactive visualizations, optimizing App performance, and effectively packaging and distributing your Apps.

Part 5: Real-World Examples provides practical demonstrations of App development through compelling examples. You will learn, by example, how to apply the concepts and techniques from the book to address real-world challenges.

The MATLAB App, codes, and examples used in the textbook will be made available via the Springer GitHub repository.

After completing reading this book, you will have a solid understanding of MATLAB App Designer and the skills to create powerful and user-friendly applications that can streamline your workflows, enhance your productivity, and effectively communicate your results.

References

1. Attaway DC. Matlab: a practical introduction to programming and problem solving. 6th ed. Butterworth-Heinemann; 2022.
2. Valentine DT, Hahn BH. Essential MATLAB for engineers and scientists. Academic Press; 2022.
3. C. Lopes P, L. Rangel R, Martha LF. An interactive user interface for a structural analysis software using computer graphics techniques in MATLAB. Computer Applications in Engineering Education. 2021;29(6):1505–25.
4. Parekh R. Fundamentals of image, audio, and video processing using Matlab®: with applications to pattern recognition. CRC Press; 2021.
5. Hossain E. App Designer and Graphical User Interface in MATLAB. MATLAB and Simulink Crash Course for Engineers. Springer; 2022. p. 299–316.
6. Lent CS. Learning to program with MATLAB: Building GUI tools. John Wiley & Sons; 2022.
7. Mathworks: MATLAB App Designer: Create desktop and web apps in MATLAB. https://au.mathworks.com/products/matlab/app-designer.html (2025). Accessed September 2025.
8. Thabit AA, Alzidi E, Mohammed MA, Palchyk P. Development of a System Based on a Graphical User Interface Using MATLAB Tools. 2024 35th Conference of Open Innovations Association (FRUCT): IEEE; 2024. p. 744–54.
9. Darweesh M. Teaching MATLAB Programming by Utilizing Image Processing and Graphical User Interface Basics. Interactive Mobile Communication, Technologies and Learning. Springer; 2021. p. 824–32.

Chapter 2
App Designer Interface

2.1 Introduction

The MATLAB App Designer interface is a powerful environment for creating interactive applications. It seamlessly integrates visual design with MATLAB's computational capabilities, making it easy to build sophisticated Apps with minimal coding effort [1, 2].

App Designer inspires anyone to create professional applications, even without extensive software development experience. You can design your Graphical User Interface (GUI) visually by dragging and dropping components, such as buttons, sliders, and plots [3, 4]. Then use the integrated editor to quickly program your App's behavior. In this chapter, we explore the key components of this interface in detail.

2.2 Canvas

The MATLAB App Designer canvas is the visual workspace where you drag and drop User Interface (UI) components to design your App's GUI layout. It provides alignment hints and allows you to intuitively arrange and size elements, seeing the immediate visual result of your design choices [3]. MATLAB App Designer canvas has the following characteristics:

- **The Heart of Design**: The central workspace where you visually construct your App's UI.
- **Drag and Drop**: You drag and drop UI components from the Component Library onto the Canvas to build your App's layout.

Supplementary Information The online version contains supplementary material available at https://doi.org/10.1007/978-981-95-8797-1_2.

G. R. Naik, *MATLAB App Development*, https://doi.org/10.1007/978-981-95-8797-1_2

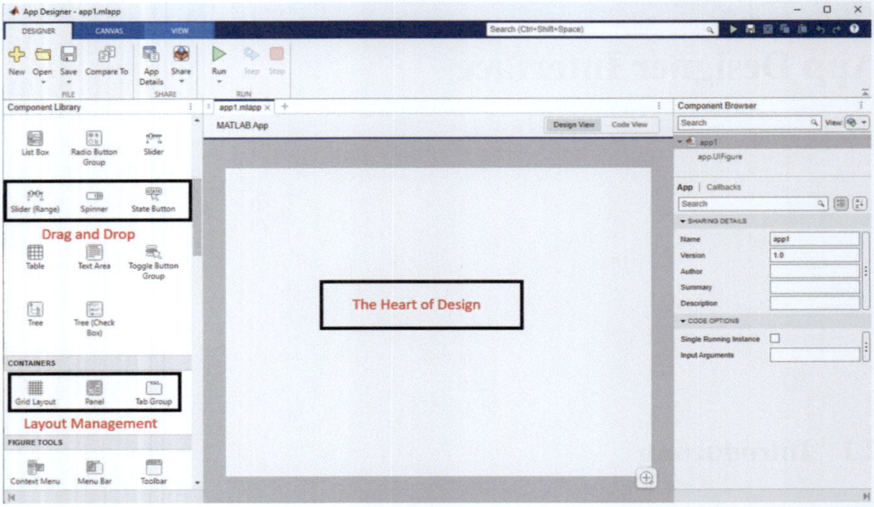

Fig. 2.1 MATLAB App Canvas layout

- **Layout Management**: Tools are provided to arrange components using various layout managers (e.g., Grid Layout) for efficient and responsive designs.
- **Visual Feedback**: As you design, Canvas provides real-time visual feedback, allowing you to see how your App will look and behave (refer to Fig. 2.1).

Here, we provide a simple example to illustrate the MATLAB App Designer canvas for an image-processing App.

- On the Canvas, you would drag and drop the following components:
  - **Image Viewer**: A UI access to display the input and processed images.
  - **File Selection Button**: To allow users to choose an image file.
  - **Sliders**: To adjust image brightness and contrast.
  - **UI Plot**: To plot an image for brightness and contrast adjustments.
- You would then use the Grid Layout Manager to arrange these components in a visually appealing and user-friendly manner (refer to Fig. 2.2). We will complete the example in detail at the end of this chapter.

2.3 Component Library

Component library is a repository of UI Elements. It consists of a vast collection of pre-built UI components, including

- **Basic Controls**: Buttons, text boxes, sliders, drop-down lists, radio buttons, and checkboxes.

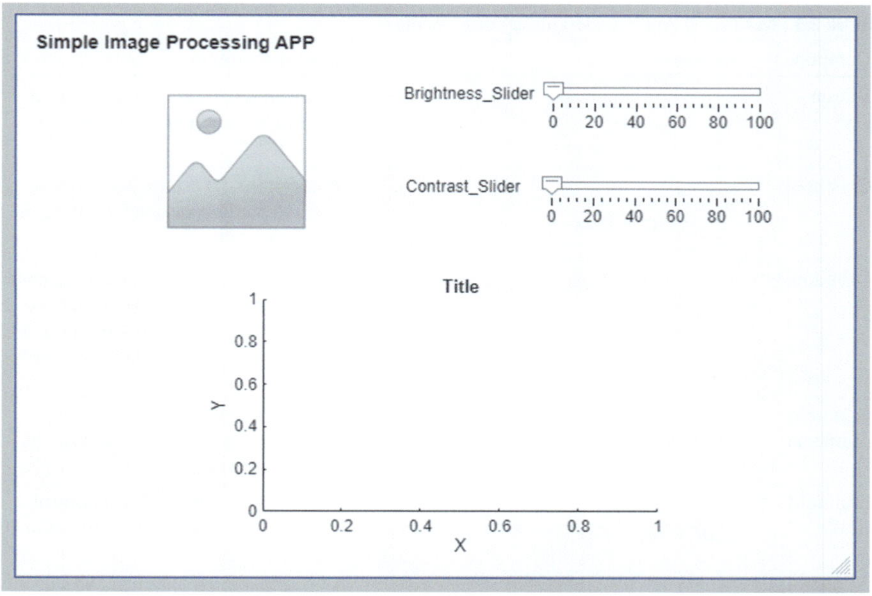

Fig. 2.2 MATLAB App Canvas layout component example

- **Data Display**: Plots, tables, gauges, and images.
- **Containers**: Panels, tabs, splitters for organizing components.
- **Drag-and-Drop Simplicity**: Easily drag and drop components from the library onto the Canvas to incorporate them into your App's design [5, 6].

Below, in Table 2.1, we outline the commonly used components and brief descriptions (we will cover some of their applications in detail with examples in Chap. 3).

Example:

- Suppose you need to allow users to select a color for a visualization. In that case, you will drag and drop a "Color Picker" component from the Component Library onto the Canvas (refer to Fig. 2.3).

Table 2.1 MATLAB App's widely used components

Component	Example	Brief description
Button		A button UI component executes a predefined function when the user clicks it
Checkbox		A checkbox UI component allows users to confirm or deny a setting or choice visually
Dropdown	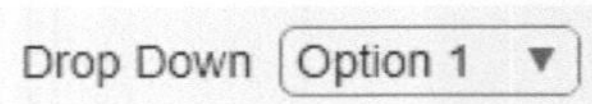	A drop-down UI component presents a list of predefined choices that can be expanded and collapsed, allowing users to select an option or enter text manually
Edit field (number)	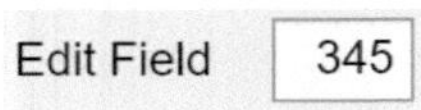	A numeric edit field provides a dedicated space for users to type in numerical data
Edit field (text)		An edit field is a text input control that allows users to enter and edit text
Date picker	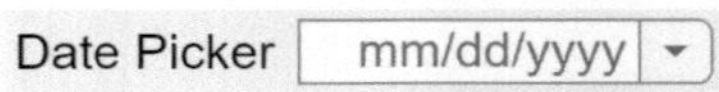	A date picker UI component enables users to select a specific date via an interactive calendar. Utilize the Date Picker object to customize the visual presentation and interactive behavior of a date picker after its creation
Text area		Allow users to input text or numerical values
Image	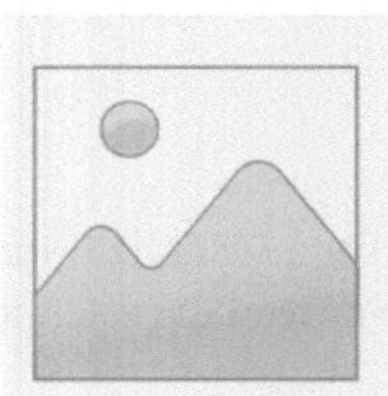	An image UI component displays a visual element, such as an icon or logo, within an application. Use the Image object to customize the appearance and interactive behavior of an image after its creation
Label	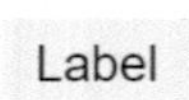	A label is a static text control that renders text on the screen

(continued)

Table 2.1 (continued)

Component	Example	Brief description
List box	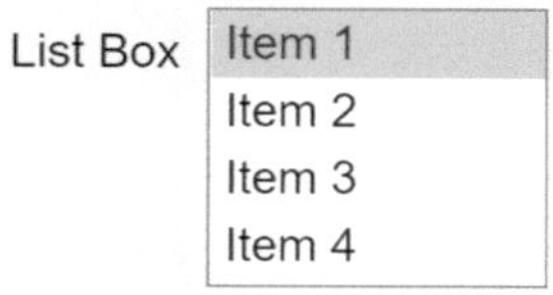	A list box allows an app user to select an option from a list. Use the ListBox object to modify the appearance and behavior of a list box after its creation
Radio buttons	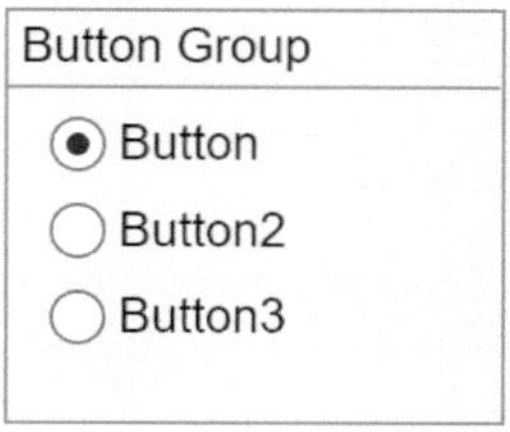	Allow users to select one option from a group
Spinners	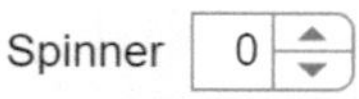	A spinner UI component enables users to choose a numerical value from a predefined range. Utilize the Spinner object to customize the visual presentation and interactive behavior of a spinner after its creation
Sliders	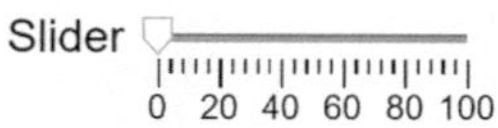	Enable users to select values from a continuous range
Toggle button	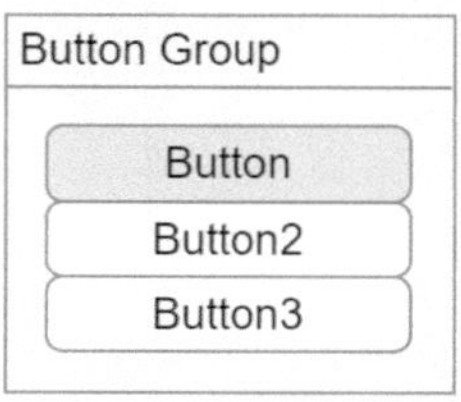	Enable users to choose a single option from a set of related choices within a button group. Utilize the Toggle Button object to customize the visual presentation and interactive behavior of an individual toggle button after its creation
UIAxes	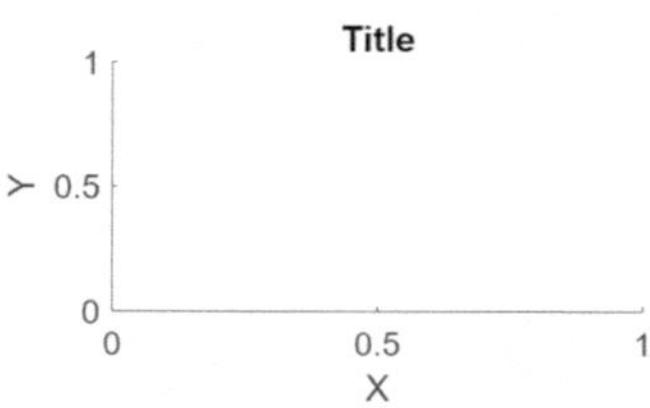	The UIAxes function creates UI axes within a new figure window and returns a UIAxes object. This function utilizes the *uifigure* function to generate the figure window

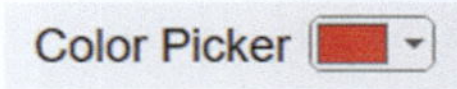

Fig. 2.3 MATLAB App Color picker example

2.4 Inspector

The MATLAB App's Inspector panel allows users to view and modify the properties and callbacks of selected UI components (such as buttons, text boxes, and plots) within the App's layout. The inspector has the following properties:

- **Customization**: Adjust the appearance (color, font), behavior (callback functions), and other properties to tailor the component to your specific needs.
- **Real-Time Updates**: Changes made in the Inspector are reflected immediately on the Canvas, providing instant visual feedback.

Example: The following MATLAB App (code) explains the usage of Inspector and its functions. The design and implementation process are as follows:

1. **Create a Figure and Plot**:
   - `figure()` creates a new figure window.
   - `plot()` generates a simple line plot with random data.
   - `title()`, `xlabel()`, and `ylabel()` add labels to the plot.
2. **Display a Message:**
   - `msgbox()` displays a message encouraging the user to select the plot line and use the Inspector:

```
function inspectorDemo()
% Create a simple figure with a plot
   fig = figure('Name', 'Inspector Demo', 'Position',
[100 100 400 300]);
   plot(1:10, rand(1,10), 'o-');

   % Add a title and labels
   title('Random Data');
   xlabel('X-axis');
   ylabel('Y-axis');

% Display a message encouraging the use of the Inspector
   msgbox('Select the plot line and use the Inspector to
explore its properties.', 'Inspector Demo');

   end
```

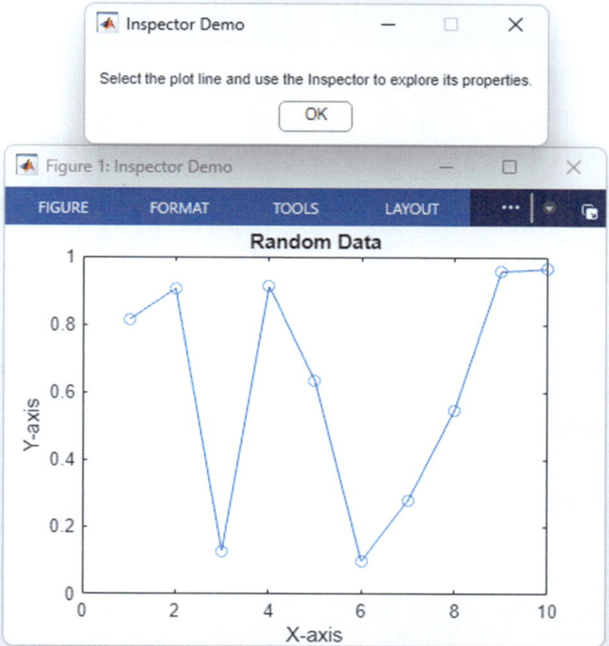

Fig. 2.4 MATLAB App Inspector example

3. **To Use the App**:
   - Save the code as an M-file (e.g., **inspectorDemo.m**).
   - Run the script in MATLAB by typing "**inspectorDemo**" in the command window or using the Run icon, which is located under the MATLAB Editor.
   - Upon executing the code, we obtain the following plot (refer to Fig. 2.4) and a message box.

2.5 Code Editor

MATLAB App's Code Editor is considered the Brain of the App, where you write the MATLAB code that defines the App's behavior. It has the following components, which include

- **Callback Functions**: Create functions that are triggered by user interactions (e.g., button clicks, slider movements).
- **App Logic**: Implement the core logic of your App, including data processing, calculations, and updates to the UI.

- **Code Completion and Debugging**: MATLAB's built-in code editor offers features such as code completion, syntax highlighting, and debugging tools to support development.

Example: The following MATLAB App explains the code editor and its functions. The design and implementation process are as follows:

1. **Create the UI**:

   - `uifigure`: Creates the main figure window for the App.
   - `uipanel`: Creates two panels: one for the code editor and one for the explanation text.
   - `uieditor`: Creates the code editor component within the code editor panel.
   - `uitextarea`: Creates a text area for displaying the explanation within the explanation panel.
   - `uibutton`: Creates a button to run the code:

```
% Create the UI
app = uifigure('Name', 'MATLAB Code Editor Explained');

% Code Editor Panel
codeEditorPanel  =  uipanel(app,  'Title',  'Code  Editor',
'Position', [20 20 400 300]);
codeEditorTable = uitable(codeEditorPanel, ...
'Position', [20 20 360 260]);
```

2. **Code Editor Initialization**:

   - Sets the initial code in the `codeEditor` with example MATLAB code:

```
codeEditorTable.Data = {
'% This is a simple MATLAB script',
'clear all; close all; clc;',
'',
'% Define variables',
'x = 1:10;',
'y = x.^2;',
'',
'% Plot the results',
'plot(x, y);',
'xlabel(''x-axis'');',
'ylabel(''y-axis'');',
'title(''Simple Plot'');'};
```

3. **Explanation Text**:

- Sets the initial text in the `explanationText` with a brief description of the code editor's key features:

```
% Explanation Panel
explanationPanel = uipanel(app, 'Title', 'Explana
tion', 'Position', [440 20 400 300]);
explanationText = uitextarea(explanationPanel, ...
    'Position', [20 20 360 260]); % Specify numer-
ic
position
explanationText.Value = ...
      ['**Key Features:**', newline, ...
       '- **Syntax Highlighting:** Colors keywords
(like "for", "if", "end"), operators, and com-
ments.', newline, ...
       '- **Code Completion:** Autocompletes func-
tion names, variable names, and built-in func-
tions.', newline, ...
       '- **Debugging Tools:** Set breakpoints, step
through code, and inspect variables.', newline, ...
       '- **Error Checking:** Identifies syntax er-
rors and provides helpful suggestions.', newline,
...
       '- **Code Folding:** Collapse sections of
code to improve readability.', newline, ...
       '- **Customizable:** Adjust font size, color
schemes, and other preferences.'];
```

4. **Run Button Functionality**:

- `runButton` uses a UI button to fetch the data from the code editor table.
- `runCode` function uses eval to execute the code from the codeEditor. It includes a try-catch block to handle potential errors during code execution and display an error message using *errordlg*:

```
% Run Button
  runButton = uibutton(app, 'Text', 'Run Code', 'Po-
sition', [20 350 100 30], ...
      'ButtonPushedFcn',       @(btn,event)      run-
Code(codeEditorTable));

  % Function to run the code
  function runCode(codeEditorTable)
      try
          % Extract code from the table
          code = string(codeEditorTable.Data);
          code = strjoin(code, newline);

          eval(code);
      catch ME
          errordlg(ME.message, 'Error');
      end
  end
```

The complete function (including the block codes described above) is given as follows:

Note: MATLAB App Designer automatically generates the required *function* and *end* code skeletons. Please paste the remaining code into the appropriate sections within these generated functions.

```
function matlabCodeEditorApp
  % Create the UI
  app  =  uifigure('Name',  'MATLAB  Code  Editor  Ex-
plained');

  % Code Editor Panel
  codeEditorPanel = uipanel(app, 'Title', 'Code Edi-
tor', 'Position', [20 20 400 300]);
  codeEditorTable = uitable(codeEditorPanel, ...
      'Position', [20 20 360 260]);

  codeEditorTable.Data = {
      '% This is a simple MATLAB script',
      'clear all; close all; clc;',
      '',
      '% Define variables',
      'x = 1:10;',
      'y = x.^2;',
      '',
      '% Plot the results',
      'plot(x, y);',
      'xlabel(''x-axis'');',
      'ylabel(''y-axis'');',
      'title(''Simple Plot'');'};
```

```
    % Explanation Panel
    explanationPanel = uipanel(app, 'Title', 'Explana-
tion', 'Position', [440 20 400 300]);
    explanationText = uitextarea(explanationPanel, ...
        'Position', [20 20 360 260]); % Specify numer-
ic position
    explanationText.Value = ...
        ['**Key Features:**', newline, ...
         '- **Syntax Highlighting:** Colors keywords
(like "for", "if", "end"), operators, and com-
ments.', newline, ...
         '- **Code Completion:** Autocompletes func-
tion names, variable names, and built-in func-
tions.', newline, ...
         '- **Debugging Tools:** Set breakpoints, step
through code, and inspect variables.', newline, ...
         '- **Error Checking:** Identifies syntax er-
rors and provides helpful suggestions.', newline,
...
         '- **Code Folding:** Collapse sections of
code to improve readability.', newline, ...
                 '- **Customizable:** Adjust font size, color
        schemes, and other preferences.'];

            % Run Button
            runButton = uibutton(app, 'Text', 'Run Code', 'Po-
        sition', [20 350 100 30], ...
                'ButtonPushedFcn',      @(btn,event)      run-
        Code(codeEditorTable));

            % Function to run the code
            function runCode(codeEditorTable)
                try
                    % Extract code from the table
                    code = string(codeEditorTable.Data);
                    code = strjoin(code, newline);

                    eval(code);
                catch ME
                    errordlg(ME.message, 'Error');
                end
            end
            end
```

After executing the app and clicking the "**Run Code**" button, we obtain the plot as shown in Fig. 2.5.

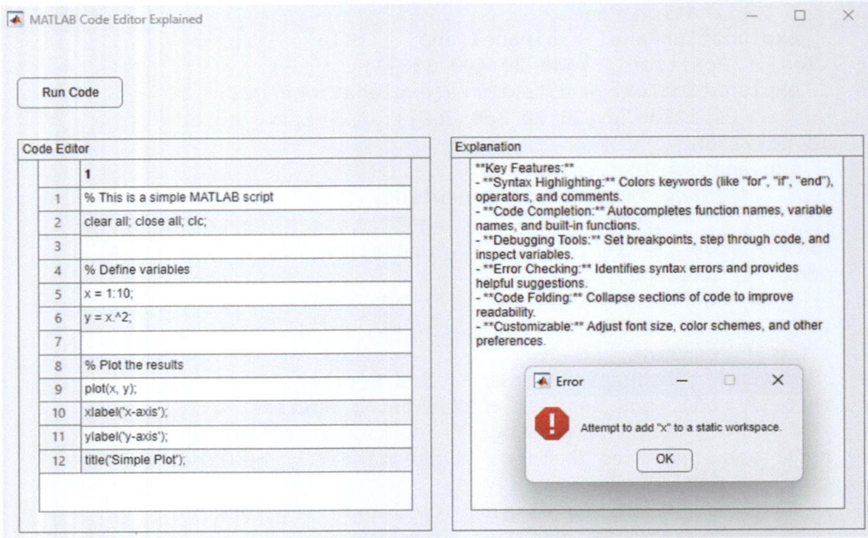

Fig. 2.5 MATLAB App Code Editor example

This App provides a foundation for exploring and understanding the MATLAB code editor. You can customize and expand it to suit your specific learning or teaching needs.

2.6 Toolstrip

The MATLAB Toolstrip is a key component of the MATLAB App desktop environment, located at the top of the window beneath the title bar. It organizes MATLAB functionality into tabs, including Home, Plots, and Apps [5, 7, 8]. In App Designer, it contains buttons and menus for common actions, such as

- Running the App
- Switching between Design and Code views
- Accessing layout tools
- Managing App properties.

Example: The following example explains Toolstrips and their usage. The design and implementation process are as follows:

1. **Create a Figure:**
   - `figure()` creates a new figure window:

```
% Create a figure
fig = figure('Name', 'Toolstrip Demo', 'Position',
[100 100 200 100]);
```

2. **Create a Toolstrip:**
   - `uitoolbar()` creates a toolstrip object and makes it a child of the figure:

```
% Create a toolstrip
ts = uitoolbar('Parent', fig);
```

3. **Add Buttons to the Toolstrip:**
   - `uipushtool()` creates push buttons and adds them to the toolstrip.
     - `Text` property sets the text displayed on the button.
     - `TooltipString` sets the text that appears when the mouse hovers over the button:

```
% Create text images (replace with your preferred
font and size)
btn1Text = imread('button1_text.png');
btn2Text = imread('button2_text.png');
btn3Text = imread('button3_text.png');

% Add buttons to the toolstrip
btn1 = uipushtool('Parent', ts, 'CData', btn1Text,
'TooltipString', 'Button 1 Action');
btn2 = uipushtool('Parent', ts, 'CData', btn2Text,
'TooltipString', 'Button 2 Action');
btn3 = uipushtool('Parent', ts, 'CData', btn3Text,
'TooltipString', 'Button 3 Action');
```

4. **Set Button Callbacks**:

   - `set(btn1,'ClickedCallback', @button1Callback)` assigns a callback function to the button.

     – The callback function is executed when the button is clicked:

```
% Set button callbacks (replace with your desired
actions)
set(btn1, 'ClickedCallback', @button1Callback);
set(btn2, 'ClickedCallback', @button2Callback);
set(btn3, 'ClickedCallback', @button3Callback);
end
```

5. **Create Callback Functions:**

   - The callback functions `(e.g., button1Callback)` define the actions to be performed when the corresponding button is clicked. In this example, they display a message in the command window:

```
% Example callback functions
function button1Callback(~, ~)
    disp('Button 1 clicked!');
end

function button2Callback(~, ~)
    disp('Button 2 clicked!');
end

function button3Callback(~, ~)
    disp('Button 3 clicked!');
end
```

The complete function (including the block codes described above) is given as follows:

```
function toolstripDemo()
% Create a figure
fig = figure('Name', 'Toolstrip Demo', 'Position',
[100 100 200 100]);
```

```
% Create a toolstrip
ts = uitoolbar('Parent', fig);

% Create text images (replace with your preferred
font and size)
btn1Text = imread('button1_text.png');
btn2Text = imread('button2_text.png');
btn3Text = imread('button3_text.png');

% Add buttons to the toolstrip
btn1 = uipushtool('Parent', ts, 'CData', btn1Text,
'TooltipString', 'Button 1 Action');
btn2 = uipushtool('Parent', ts, 'CData', btn2Text,
'TooltipString', 'Button 2 Action');
btn3 = uipushtool('Parent', ts, 'CData', btn3Text,
'TooltipString', 'Button 3 Action');

% Set button callbacks (replace with your desired
actions)
set(btn1, 'ClickedCallback', @button1Callback);
set(btn2, 'ClickedCallback', @button2Callback);
set(btn3, 'ClickedCallback', @button3Callback);
end

% Example callback functions
function button1Callback(~, ~)
    disp('Button 1 clicked!');
end

function button2Callback(~, ~)
    disp('Button 2 clicked!');
end

function button3Callback(~, ~)
    disp('Button 3 clicked!');
end
```

To Use the App:

- Save the code: Save the code as an M-file (such as **toolstripDemo.m**).
- Run the script: Type "**toolstripDemo**" in the MATLAB command window or use the Run icon, located under the MATLAB Editor.

Upon clicking on each toolstrip (sub-image), it displays the following in the MATLAB workspace:

>> toolstripDemo
Button 1 clicked!
Button 2 clicked!
Button 3 clicked!

This example demonstrates a simple toolstrip with three buttons (refer to Fig. 2.6).

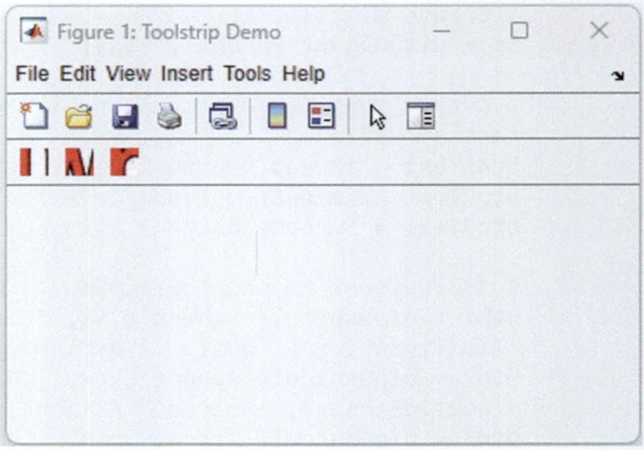

Fig. 2.6 MATLAB App Toolstrip example

By effectively utilizing the components and tools explained above, you can design and develop a wide range of interactive MATLAB applications, from simple data analysis tools to complex scientific simulations.

2.7 Simple Image Processing App Example

Here, we implement a simple image-processing app discussed in Sect. 2.2. The detailed steps in designing the App are outlined as follows:

- **Step 1**: Define App properties

Click on Code View, below your component properties, and add the following code:

lowing code:

```
properties (Access = private)
        OriginalImage % this is to store the original image
        ProcessedImage % this is to store the processed im-
age, after slider functions
    end
```

- Step 2: Add the image button callback (browse and load image)

Click on your Image Button in Design View. In the Inspector, scroll to Callbacks → + ImageClickedFcn. This callback function works as follows: when you click the image icon, it opens a file browser, loads the selected image, and displays it in the UIAxes. Add the following MATLAB function:

Note: MATLAB App Designer automatically generates the required *function* and *end* code skeletons. Please paste the remaining code into the appropriate sections within these generated functions.

```
% Image clicked function: Image
        function ImageClicked(app, event)
            [file, path] = uigetfile({'*.jpg;*.png;*.bmp'},
'Select an Image');
            if isequal(file,0)
                return;
            end
            img = imread(fullfile(path, file));
            app.OriginalImage = img;
            app.ProcessedImage = img;
            imshow(img, 'Parent', app.ImageAxes);
        end
```

- Step 3: Add sliders callback (Brightness and Contrast)

Now you will learn how to modify the displayed image dynamically using your sliders. First, click your Brightness_Slider. In Callbacks, add + ValueChangedFcn.

Make sure to set the brightness in the range [−0.5, 0.5] and the contrast values ranging [0.5, 2].

Add the following MATLAB function:

```
% Value changed function: Brightness_Slider
        function Brightness_SliderValueChanged(app, event)
            updateImage(app);
        end
```

Similarly, click on your Contrast_Slider. In Callback, add Add + ValueChangedFcn and add the following MATLAB code:

```
% Value changed function: Contrast_Slider
        function Contrast_SliderValueChanged(app, event)
            updateImage(app);
        end
```

- Step 4: Add an Image processing function

Now, at the bottom of your code (make sure it is still inside the **classdef**), add the following function:

```
function updateImage(app)
        if isempty(app.ProcessedImage)
            return;
        end
    brightness = app.Brightness_Slider.Value;
    contrast = app.Contrast_Slider.Value;
    img = im2double(app.ProcessedImage);
    img = img * contrast + brightness;
    img = max(min(img,1),0); % clip between 0-1

    imshow(img','Parent', app.ImageAxes);
end
```

- Step 4: Save and Run the MATLAB App

Now, save your completed App as "**Simple_Image_Processing_App**". Run the MATLAB App, click the **image button** → choose an image. Image appears on **UIAxes**. Move sliders → the displayed image updates live with new brightness and contrast (refer to Fig. 2.7).

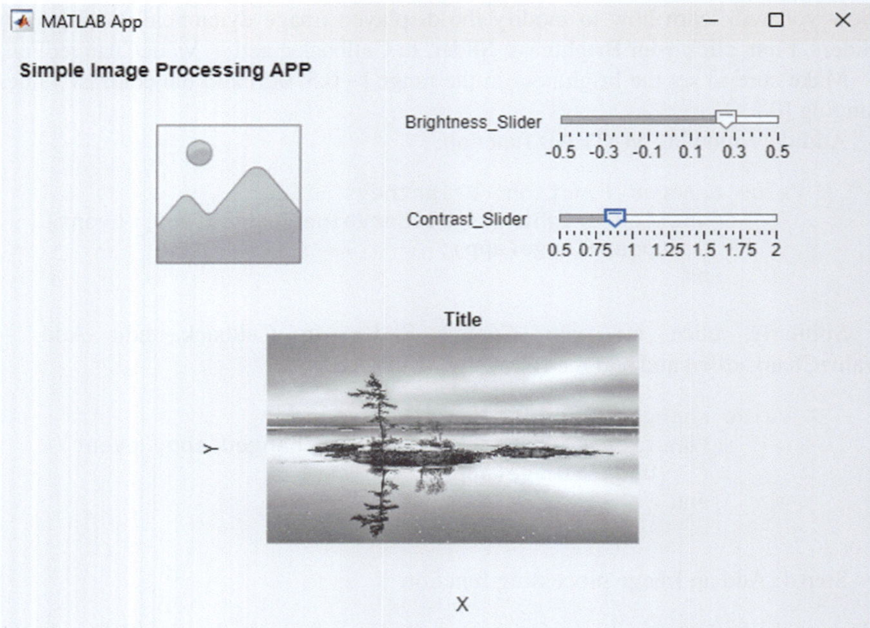

Fig. 2.7 Simple Image Processing MATLAB App example

2.8 Conclusion

The App Designer's graphics interface, comprising the Canvas, Component Library, Inspector, Code Editor, and Toolstrip, provides the efficiency needed to create sophisticated MATLAB applications. This chapter provided the reader with example applications that demonstrate the use of the five components mentioned above. By integrating graphical interfaces with powerful computational tools, App Designer provides the efficiency users need to develop applications. By facilitating prototyping and redesigning and ultimately speeding up the creation of high-quality MATLAB applications, App Designer provides an efficient graphical interface.

References

1. Attaway DC. Matlab: a practical introduction to programming and problem solving. 6th ed. Butterworth-Heinemann; 2022.
2. Lent CS. Learning to program with MATLAB: Building GUI tools. John Wiley & Sons; 2022.
3. Mathworks: MATLAB App Designer: Create desktop and web apps in MATLAB. https://au.mathworks.com/products/matlab/app-designer.html (2025). Accessed September 2025.
4. Hossain E. App Designer and Graphical User Interface in MATLAB. MATLAB and Simulink Crash Course for Engineers. Springer; 2022. p. 299–316.
5. Akan A, Chaparro LF. Signals and systems using MATLAB®. Academic Press; 2024.
6. Schepers J, Hofmans J. TwoMP: A MATLAB graphical user interface for two-mode partitioning. Behavior research methods. 2009;41(2):507–14.
7. C. Lopes P, L. Rangel R, Martha LF. An interactive user interface for a structural analysis software using computer graphics techniques in MATLAB. Computer Applications in Engineering Education. 2021;29(6):1505–25.
8. Parekh R. Fundamentals of image, audio, and video processing using Matlab®: with applications to pattern recognition. CRC Press; 2021.

Part II
App Designer Fundamentals

Chapter 3
Core User Interface Components

3.1 Introduction

The core User Interface (UI) components of MATLAB App Designer are the fundamental building blocks for creating interactive Graphical User Interfaces (GUIs). These pre-built elements allow you to design the visual layout and user interaction points of your application without writing any low-level code (graphics code). They range from basic input and display elements to more complex interactive controls, enabling programmers/developers to assemble functional, visually appealing Apps quickly [1–3].

In the previous chapter (Chap. 2), we briefly outlined different UI components. This chapter demonstrates the functions and usage of Core UI components used for building basic MATLAB Apps using App Designer. Below, we explain how these UI components are used to create sophisticated, interactive applications.

3.2 Basic Components

Buttons: In the MATLAB App, buttons trigger actions (e.g., starting a calculation, loading data, or exiting an application). **Example**: A "Calculate" button that triggers a function to perform a mathematical operation is shown in Fig. 3.1.

Text Boxes: In the MATLAB App, text boxes permit users to input text or numerical values. **Example**: Text boxes for entering input values for a calculation or specifying file paths are shown in Fig. 3.2.

Supplementary Information The online version contains supplementary material available at https://doi.org/10.1007/978-981-95-8797-1_3.

G. R. Naik, *MATLAB App Development*, https://doi.org/10.1007/978-981-95-8797-1_3

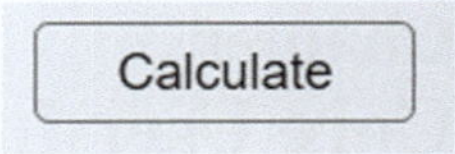

Fig. 3.1 An example of a Button

Fig. 3.2 An example of a Text box

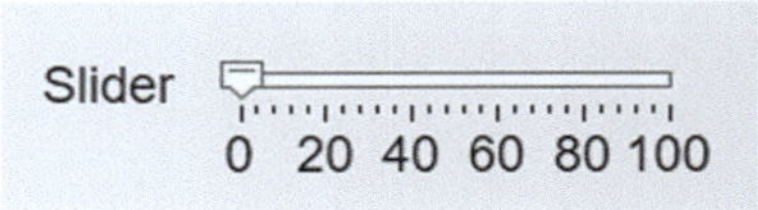

Fig. 3.3 An example of a Slider

Sliders: In the MATLAB App, sliders enable users to select values from a continuous range. **Example**: A slider to adjust the brightness of an image is shown in Fig. 3.3.

3.3 Dropdown List

A dropdown list is a UI component that presents a list of choices to the user, allowing them to select one option from a predefined set. It is a compact way to offer multiple selections without taking up much screen space until activated. A dropdown list provides a list of options for users to choose from [4].

Here, we explain how to create a standard MATLAB App that allows the user to select an item from a list and then displays the selection directly in the window. The process is as follows:

- **Define the Main Function**: You start by creating a MATLAB function (e.g., **matlab_dropdown_app.m**). This function will contain all the code for your app:

```
function matlab_dropdown_app()

% MATLAB_DROPDOWN_APP creates a simple MATLAB app with a
dropdown list
% and a text area that displays the selected option.

End
```

Note: MATLAB App Designer automatically generates the required *function* and *end* code skeletons. Please paste the remaining code into the appropriate sections within these generated functions.

- **Set Up the App Window** (`uifigure`)**:** Inside your function, use uifigure to create the main window where all your app's components will live. You will specify its size and position:

```
% 1. Create the main figure (app window)
% uifigure is the recommended type for modern MATLAB Apps.
app.UIFigure = uifigure('Name', 'MATLAB Dropdown Selector',
...
                        'Position', [100 100 350 350], ...
% Increased figure height to 350
                        'Resize', 'off'); % Prevent resiz-
ing for simplicity
```

- **List Your Dropdown Options**: Create a cell array (such as app.Options) to hold all the text strings that will appear as choices in your dropdown list:

```
% 2. Define the options for the dropdown list
% These will be the 'Items' property of the uilistbox.
app.Options = {'Choose an option...', 'Apple', 'Banana',
'Orange', 'Grape'};
```

- **Create the Dropdown List** (`uilistbox`)**:** Use `uilistbox` to place the dropdown component onto your app window. You will mention which items (your options) to display and set their initial value:

```
% 3. Create the dropdown list (uilistbox)
% uilistbox is used for single- or multiple-selection
lists.
% Adjusted position to accommodate new label positions.
app.Dropdown = uilistbox(app.UIFigure, ...
                        'Items', app.Options, ... % Set
the options
                        'Position', [20 70 150 200], ... %
[left bottom width height]
                        'Value', app.Options{1}, ... % Set
initial selected item (the placeholder)
                        'ValueChangedFcn', @(src, event)
dropdownSelectionChanged(app, src, event)); % Callback
function
```

- **Add a Text Display Area** (`uitextarea`)**:** Include a `uitextarea` where messages and the user's selections can be shown. Ensure that it is set to "**off**" for Editable, so users can't type in it.

```
% 4. Create a text area to display messages and selections
% uitextarea is used for multi-line text display.
% Adjusted position to accommodate new label positions.
app.MessageArea = uitextarea(app.UIFigure, ...
                      'Position', [180 70 150 200], ...
                             'Editable', 'off', ... % Make
it read-only
                      'FontColor', [0 0 0.5], ... % Dark
blue text
                             'BackgroundColor', [0.95 0.95
1]); % Light blue background
```

- **Add Labels for Clarity** (`uilabel`)**:** Use `uilabel` to place titles (such as "Interactive Dropdown Selector") and prompts (like "Select your favorite fruit:") in your app, ensuring they're positioned clearly and don't overlap:

```
% 5. Create a title label for the app
% Adjusted position to move it higher, preventing overlap.
uilabel(app.UIFigure, ...
        'Text', 'Interactive Dropdown Selector', ...
        'Position', [20 300 310 22], ... % Moved higher
        'HorizontalAlignment', 'center', ...
        'FontWeight', 'bold', ...
        'FontSize', 16);

% 6. Create a label for the dropdown
% Adjusted position to move it higher, just above the
dropdown.
uilabel(app.UIFigure, ...
        'Text', 'Select your favorite fruit:', ...
        'Position', [20 270 150 22], ... % Moved higher
        'HorizontalAlignment', 'left', ...
        'FontWeight', 'bold');
```

- **Set Initial Message**: Provide a default message in your `uitextarea` when the app first opens, guiding the user:

```
% 7. Set an initial message in the text area
app.MessageArea.Value = 'Please select an option from the
dropdown.';
```

- Define the Callback Function (dropdownSelectionChanged): Create a nested function (e.g., dropdownSelectionChanged) that will automatically run whenever the user selects a new item from the dropdown.
- Implement Callback Logic: Within this callback, retrieve the selected option from the dropdown. Then, based on the selection, update the text in your `uitextarea` to provide instant feedback to the user:

```
% --- Callback function for dropdown selection change ---
% This function is executed whenever the user selects a new
item in the dropdown.
function dropdownSelectionChanged(app, src, event)
    % Get the currently selected item's string value.
    % For uilistbox with single selection, 'Value' directly
returns the string.
    selectedOption = src.Value;

    % Clear previous messages in the text area
    app.MessageArea.Value = '';

    % Update the message based on the selection
    if strcmp(selectedOption, app.Options{1}) % Check if
the placeholder is selected
        app.MessageArea.Value = 'Please select an option
from the dropdown.';
    else
        app.MessageArea.Value = ['You selected: ', se-
lectedOption, '!'];
    end

    % Display raw value for debugging/clarity (optional,
prints to Command Window)
    disp(['Raw value selected: ', selectedOption]);
end
```

The whole function of the "**matlab_dropdown_app**" is given as follows:

Note: MATLAB App Designer automatically generates the required *function* and *end* code skeletons. Please paste the remaining code into the appropriate sections within these generated functions.

```
function matlab_dropdown_app()

% MATLAB_DROPDOWN_APP creates a simple MATLAB app with a
dropdown list
% and a text area that displays the selected option.

% 1. Create the main figure (app window)
% uifigure is the recommended type for modern MATLAB apps.
app.UIFigure = uifigure('Name', 'MATLAB Dropdown Selector',
...
                        'Position', [100 100 350 350], ...
% Increased figure height to 350
                        'Resize', 'off'); % Prevent resiz-
ing for simplicity
```

```
% 2. Define the options for the dropdown list
% These will be the 'Items' property of the uilistbox.
app.Options = {'Choose an option...', 'Apple', 'Banana',
'Orange', 'Grape'};

% 3. Create the dropdown list (uilistbox)
% uilistbox is used for single or multiple selection lists.
% Adjusted position to accommodate new label positions.
app.Dropdown = uilistbox(app.UIFigure, ...
                         'Items', app.Options, ... % Set
the options
                         'Position', [20 70 150 200], ... %
[left bottom width height]
                         'Value', app.Options{1}, ... % Set
initial selected item (the placeholder)
                         'ValueChangedFcn', @(src, event)
dropdownSelectionChanged(app, src, event)); % Callback
 function

 % 4. Create a text area to display messages and selections
 % uitextarea is used for multi   -line text display.
 % Adjusted position to accommodate new label positions.
 app.MessageArea = uitextarea(app.UIFigure,     ...
                         'Position' , [180 70 150 200],   ...
                                 'Editable' , 'off' , ... % Make
 it read -only
                         'FontColor' , [0 0 0.5],   ... % Dark
 blue text
                                 'BackgroundColor'  , [0.95 0.95
 1]);  % Light blue background

 % 5. Create a title label for the app
 % Adjusted position   to move it higher, preventing overlap.
 uilabel(app.UIFigure,   ...
          'Text' , 'Interactive Dropdown Selector'   , ...
          'Position' , [20 300 310 22],   ... % Moved higher
          'HorizontalAlignment'  , 'center' , ...
          'FontWeight' , 'bold' , ...
          'FontSize' , 16);

 % 6. Create a label for the dropdown
 % Adjusted position to move it higher, just above the
 dropdown.
 uilabel(app.UIFigure,   ...
```

```
'Text'   ,  'Select your favorite fruit:'           ,  ...
'Position'    , [20 270 150 22],          ...  % Moved higher
'HorizontalAlignment'         ,  'left'   ,  ...
'FontWeight'     ,  'bold'   );

% 7. Set an initial message in the text area
app.MessageArea.Value = 'Please select an option from the
dropdown.';

% --- Callback function for dropdown selection change ---
% This function is executed whenever the user selects a new
item in the dropdown.
function dropdownSelectionChanged(app, src, event)
    % Get the currently selected item's string value.
    % For uilistbox with single selection, 'Value' directly
returns the string.
    selectedOption = src.Value;

    % Clear previous messages in the text area
    app.MessageArea.Value = '';

    % Update the message based on the selection
    if strcmp(selectedOption, app.Options{1}) % Check if
the placeholder is selected
        app.MessageArea.Value = 'Please select an option
from the dropdown.';
    else
        app.MessageArea.Value = ['You selected: ', se-
lectedOption, '!'];
    end

    % Display raw value for debugging/clarity (optional,
prints to Command Window)
    disp(['Raw value selected: ', selectedOption]);
end

end
```

To use this App, open the App. After running the **matlab_dropdown_app** command in MATLAB, a new window will appear:

1. **Make a Selection**: Click on the dropdown list (it will initially show "**Choose an option**…"). A list of fruits will appear. Click on any fruit you want to select.
2. **Observe the Result**: The text area next to the dropdown will display the item (Fig. 3.4).
3. Upon clicking the selection, the following results will appear in the MATLAB workspace:

Fig. 3.4 An example of a MATLAB Interactive dropdown selector App

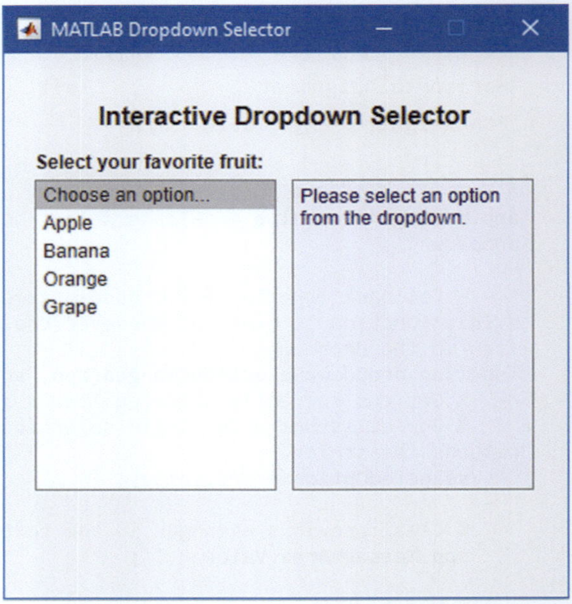


```
>> matlab_dropdown_app
Raw value selected: Banana
Raw value selected: Apple
Raw value selected: Orange
```

3.4 Radio Buttons

A radio button is a UI component that allows users to select only one option from a predefined set of mutually exclusive choices. When one radio button in a group is selected, any previously selected button in that group is automatically unselected, ensuring a single choice. Radio buttons allow users to choose one option from a group [4].

Example: Here we demonstrate a simple MATLAB App using Radio Buttons. The design and implementation process are as follows:

- Define Units: A cell array containing the available units for the chosen quantity (e.g., length):

```
% Define available units for a specific quantity (e.g.,
length)
units = {'meters', 'centimeters', 'kilometers'};
```

- Create the App using the following UI elements:

figure(): creates the main figure window of the App.
uicontrol(): creates a set of radio buttons within a loop:
Style: "radiobutton"
String: The name of the unit.
Position: Specifies the position of each radio button.
Callback: Specifies the radiobutton_callback function to be executed when a radio button is selected:

```
% Create a figure
f = figure('Position', [100 100 250 150]);

% Create radio buttons for each unit
for i = 1:length(units)
    radiobutton(i) = uicontrol('Style', 'radiobutton',
'String', units{i}, ...
       'Position', [20, 120 - (i-1)*30, 150, 20], ...
        'Callback', @radiobutton_callback);
    end
```

- Set Default Selection: **set(radiobutton(1), 'Value', 1)**: sets the first radio button as selected by default:

```
% Set the first radio button as selected by default
set(radiobutton(1), 'Value', 1);
```

- Radiobutton callback Function: **get(hObject, 'String')**: receives the text (unit name) of the selected radio button. disp(['Selected unit: ', selected_unit]): Displays the unit chosen in the command window:

```
% Function to handle radio button selection
function radiobutton_callback(hObject, eventdata)
    selected_unit = get(hObject, 'String');

% Display the selected unit (or perform further actions)
    disp(['Selected unit: ', selected_unit]);
end
```

The whole function for the Radio Button App (described above) is given as follows:

```
% Define available units for a specific quantity (e.g.,
length)
units = {'meters', 'centimeters', 'kilometers'};
% Create radio buttons for each unit
for i = 1:length(units)
    radiobutton(i) = uicontrol('Style', 'radiobutton',
'String', units{i}, ...
                              'Position', [20, 120 - (i-
1)*30, 150, 20], ...
                              'Callback',
@radiobutton_callback);
end

% Set the first radio button as selected by default
set(radiobutton(1), 'Value', 1);

% Function to handle radio button selection
function radiobutton_callback(hObject, eventdata)
    selected_unit = get(hObject, 'String');

    % Display the selected unit (or perform further ac-
tions)
    disp(['Selected unit: ', selected_unit]);
end

end
```

Please Do the Following to Use This App:

1. Save the code as an M-file (e.g., **`unit_selector_radio_app.m`**).
2. Run the script in MATLAB.
3. The app will display a set of radio buttons for the defined units.
4. Select a unit by clicking on the corresponding radio button (Fig. 3.5).
5. Upon clicking the selection, the following results will appear in the MATLAB workspace:

   >> unit_selector_radio_app
   Selected unit: centimeters (*after clicking "centimeters radio button"*)
   Selected unit: kilometers (*after clicking "kilometers radio button"*)
   Selected unit: meters (*after clicking "meters radio button"*)

3.5 Checkboxes

A checkbox is a UI component that allows a user to select one or more options from a set of independent choices. Unlike radio buttons, multiple checkboxes within a group can be selected simultaneously, as each represents a distinct, non-exclusive selection. Checkboxes allow users to choose various options from a list [4].

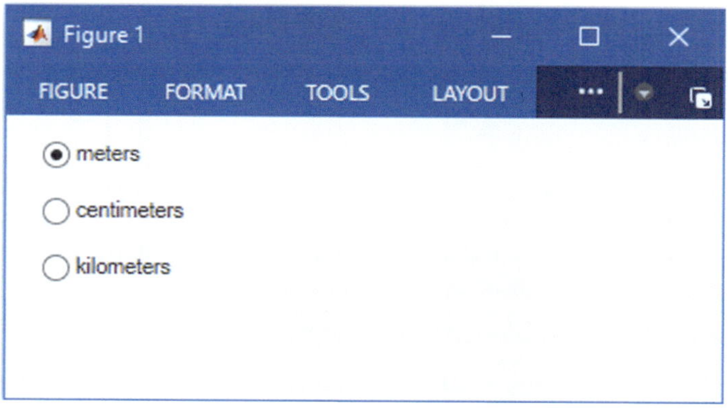

Fig. 3.5 An example of a MATLAB Radio Button App

Example: Here we demonstrate a simple MATLAB App using Checkboxes. The design and implementation process are as follows:

1. **Define Features:** A cell array containing the names of the available report features:

```
% Define available report features
       features = {'Introduction', 'Methods', 'Results',
'Discussion', 'Conclusion', 'References'};
```

2. **Create the GUI:**

- `figure()` creates the main figure window.
- `uicontrol()` creates a set of checkboxes within a loop:
  - `Style`: "checkbox"
  - `String`: The name of the feature.
  - `Position`: Specifies the position of each checkbox.
  - `Value`: 1 (default: all features are selected).
  - `uicontrol()` creates a button to trigger report generation:

```
% Create a figure
     f = figure('Position', [100 100 250 250]);

% Create checkboxes for each feature
     for i = 1:length(features)
          checkbox(i)   =   uicontrol('Style',   'checkbox',
     'String', features{i}, ...
                                  'Position', [20, 200 -
     (i-1)*25, 150, 20], ...
                                  'Value', 1); % Default:
     all features selected
     End
```

3. **`Generate_report_callback`** Function:

- `selected_features`: An empty cell array to store the names of the selected features.
- The loop iterates through each checkbox:
  - `get(checkbox(i), 'Value')` checks if the checkbox is checked (value is 1).
  - If the checkbox is checked, the corresponding feature name is added to the `selected_features` cell array.
- Displays the list of selected features in the command window.
  - This is where you would typically proceed with generating the report based on the selected features:

```
% Create a button to generate the report (example)
btn_generate = uicontrol('Style', 'pushbutton',
'String', 'Generate Report', ...
                         'Position', [60, 50, 100, 25],
'Callback', @generate_report_callback);

% Function to handle report generation
function generate_report_callback(hObject, eventdata)
    selected_features = {};
    for i = 1:length(features)
       if get(checkbox(i), 'Value') == 1
            selected_features{end+1} = features{i};
        end
    end

% Display selected features (or proceed with report gen-
eration)
          disp('Selected Features:');
      disp(selected_features);
end
```

The entire function of the Checkboxes App (described above) is given as follows:

```
function report_feature_selector_app()
  % Define available report features
  features = {'Introduction', 'Methods', 'Results',
'Discussion', 'Conclusion', 'References'};
 % Create a figure
  f = figure('Position', [100 100 250 250]);

  % Create checkboxes for each feature
  for i = 1:length(features)
      checkbox(i) = uicontrol('Style', 'checkbox',
'String', features{i}, ...
                              'Position', [20, 200 -
(i-1)*25, 150, 20], ...
                              'Value', 1); % Default:
all features selected
  end

 % Create a button to generate the report (example)
  btn_generate = uicontrol('Style', 'pushbutton',
'String', 'Generate Report', ...
                            'Position', [60, 50, 100,
25], 'Callback', @generate_report_callback);

  % Function to handle report generation
  function generate_report_callback(hObject, event-
data)
      selected_features = {};
      for i = 1:length(features)
          if get(checkbox(i), 'Value') == 1
              selected_features{end+1} = fea-
tures{i};
          end
      end
      % Display selected features (or proceed with
report generation)
      disp('Selected Features:');
  disp(selected_features);
  end
  end
```

To use the App, please do the following:

1. Save the code as an M-file (e.g., `report_feature_selector_app.m`).
2. Run the script in MATLAB.
3. The app will display a set of checkboxes for the available report features.
4. Uncheck any features you don't want to include in the report.
5. Click the "Generate Report" button.
6. The app will display the list of selected features in the command window (Fig. 3.6).

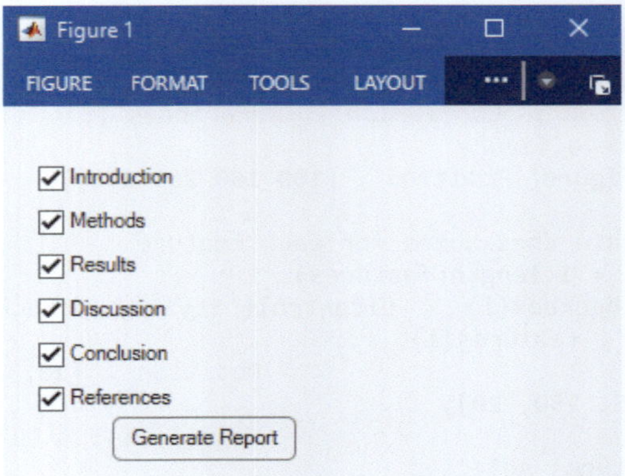

Selected Features:
{'Introduction'} {'Methods'} {'Results'} {'Discussion'} {'Conclusion'}
{'References'}

Fig. 3.6 An example of the MATLAB Checkboxes App. Selected Features: {'Introduction'} {'Methods'} {'Results'} {'Discussion'} {'Conclusion'} {'References'}

3.6 Plots

Plots in MATLAB App Designer are UI components that are specifically axes components and serve as the visual canvas for displaying graphical representations of data. They enable users to embed various types of plots (line, scatter, bar, and 3D) directly within their interactive applications, allowing them to visualize and explore data within the GUI [4, 5]. The axes component provides the framework for plotting, and you use MATLAB's standard plotting functions (e.g., `plot`, `scatter`, `bar`) to draw data onto these axes. Display data graphically (line plots, bar charts, scatter plots, etc.).

Example: Here we demonstrate a simple MATLAB plotting app. The design and implementation process are as follows:

1. **Create a Figure**:

   - `figure()` creates a new figure window:

```
% Create a figure
  fig = figure('Name', 'Simple Plot Demo', 'Position', [100
100 400 300]);
```

2. **Generate Data**:

   - **`x is an array of values from 0 to 10 with an increment of 0.1`**.
   - `y` is an array of sine values corresponding to the `x` values:

```
% Generate sample data
x = 0:0.1:10;
y = sin(x);
```

3. **Create a Plot**:

   - `plot(x, y, 'r-','LineWidth', 2)` creates a line plot:
     - `x` and `y` are the data to be plotted.
     - `'r-'` specifies a red line style.
     - `'LineWidth', 2` sets the line width to 2 points:

```
% Create a plot
plot(x, y, 'r-', 'LineWidth', 2);
```

4. **Add Labels and Title**:

   - `xlabel()`, `ylabel()`, and `title()` add labels to the axes and a title to the plot:

```
% Add labels and title
xlabel('X-axis');
ylabel('Y-axis');
title('Sine Wave');
```

5. **Add Grid Lines**:

   - `grid on` displays grid lines on the plot:

```
% Add grid lines
grid on;
```

The entire function of the Plot App (described above) is given as follows:

```
function plotDemo()

% Create a figure
fig = figure('Name', 'Simple Plot Demo', 'Position', [100
100 400 300]);

    % Generate sample data
    x = 0:0.1:10;
    y = sin(x);

    % Create a plot
    plot(x, y, 'r-', 'LineWidth', 2);

    % Add labels and title
    xlabel('X-axis');
    ylabel('Y-axis');
    title('Sine Wave');

    % Add grid lines
    grid on;

end
```

To use the App, please do the following:

1. Save the code as an M-file (e.g., `plotDemo.m`).
2. Run the script in MATLAB by typing `plotDemo` in the command window.

This will create a figure displaying a red sine wave plot with labels and grid lines (Fig. 3.7).

3.7 Tables

A table in MATLAB App Designer is a UI component that displays and allows interaction with tabular data. It presents data in a structured row-and-column format, similar to a spreadsheet, enabling users to view, sort, and sometimes edit data directly within the application, and to display it in a tabular format [4, 6].

Example: Here, we demonstrate a simple MATLAB app that explains the functionalities and usage of Tables. The design and implementation process are as follows:

1. **Create Sample Data:**

    - `Name`, `Age`, and `City` are cell arrays containing sample data:

```
% Create sample data
Name = {'Alice', 'Bob', 'Charlie', 'David'};
Age = [25, 31, 22, 28];
City = {'New York', 'London', 'Paris', 'Tokyo'};
```

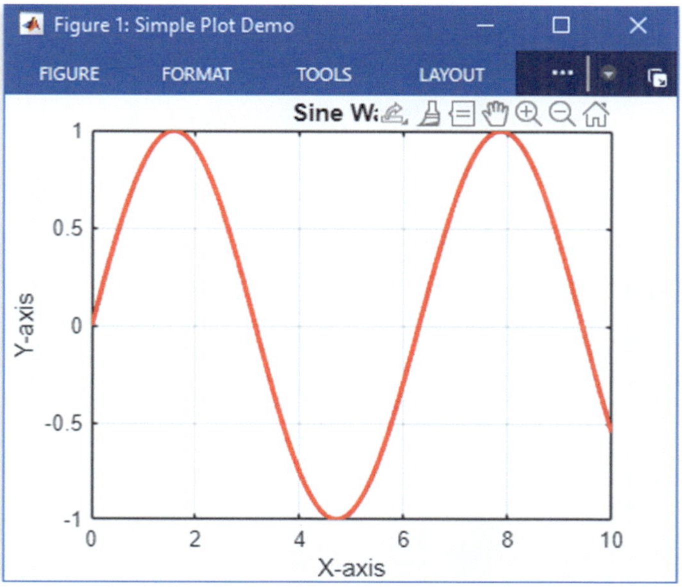

Fig. 3.7 An example of the MATLAB Plots App

2. **Create a Table:**
   - `myTable = table(Name', Age', City', 'VariableNames', {'Name', 'Age', 'City'});` creates a table named `myTable` with the specified data.
     - `'` transposes the row vectors to column vectors.
     - `'VariableNames'` specifies the column headers:

```
% Create a table
  myTable = table(Name', Age', City', 'VariableNames',
  {'Name', 'Age', 'City'});
```

3. **Display the Table**:
   - `disp(myTable)` displays the table in the command window:

```
% Display the table
disp(myTable);
```

4. **Access and Manipulate Data:**
   - `myTable{1, 'Age'}` accesses the age of the first row (Alice).
   - `myTable.Age(2) = 31` updates Bob's age in the table:

```
% Access and manipulate data
disp('Age of Alice:');
disp(myTable{1, 'Age'});

myTable.Age(2) = 31; % Update Bob's age
disp('Updated Table:');
disp(myTable);
```

The entire function of the Table App (described above) is given as follows:

```
function tableDemo()
% Create sample data
Name = {'Alice', 'Bob', 'Charlie', 'David'};
Age = [25, 31, 22, 28];
City = {'New York', 'London', 'Paris', 'Tokyo'};

% Create a table
myTable = table(Name', Age', City', 'VariableNames',
{'Name', 'Age', 'City'});

% Display the table
disp(myTable);

% Access and manipulate data
disp('Age of Alice:');
disp(myTable{1, 'Age'});

myTable.Age(2) = 31; % Update Bob's age
disp('Updated Table:');
disp(myTable);
end
```

To Use the App:

1. Save the code as an M-file (e.g., **`tableDemo.m`**).
2. Run the script in MATLAB by typing **`tableDemo`** in the command window.

This simple app demonstrates the basic creation and manipulation of a MATLAB table (Fig. 3.8).

Fig. 3.8 An example of the MATLAB Table App

```
Updated Table:
       Name        Age        City
    ___________    ___    ____________

    {'Alice'  }    25     {'New York'}
    {'Bob'    }    31     {'London'  }
    {'Charlie'}    22     {'Paris'   }
    {'David'  }    28     {'Tokyo'   }
```

3.8 UIGrid

A UI grid in MATLAB App Designer is a container component that organizes other UI components (such as buttons, text fields, and panels) into a flexible grid layout of rows and columns. It simplifies component arrangement and alignment, ensuring a clean, responsive interface that adapts well to different window sizes. Arrange components in a grid-like structure for flexible and organized layouts [7, 8].

Example: Here, we demonstrate a simple MATLAB app that explains the functionalities and usage of UIGrids. The design and implementation process are as follows:

1. **Creating the UI Figure**
   - `app = uifigure('Name', 'UIGrid Example');`
     - This line creates a new figure window for your app and gives it the title "UIGrid Example":

```
% Create the UI
app = uifigure('Name', 'UIGrid Example');
```

2. **Creating a Grid Layout**
   - `grid = uigridlayout(app, [2 2]);`
     - This creates a grid layout within the figure.

       `[2 2]` specifies that the grid will have 2 rows and 2 columns.
     - The `uigridlayout` function returns a `GridLayout` object, which is stored in the variable `grid`:

```
% Create a 2x2 grid layout
grid = uigridlayout(app, [2 2]);
% Create UI components
button1 = uibutton(grid, 'Text', 'Button 1');
button2 = uibutton(grid, 'Text', 'Button 2');
editField = uieditfield(grid, 'Value', 'Enter Text');
text = uilabel(grid, 'Text', 'Some Text');
```

3. **Creating UI Components**
   - button1 = uibutton(grid, 'Text', 'Button 1');
     - This creates a button within the grid.

`grid` specifies that the button will be placed within the grid layout.
`'Text', 'Button 1'` sets the text displayed on the button.

- Similarly, `button2`, `editField`, and `text` create a second button, an editable text field, and a label within the grid:

```
% Create UI components
button1 = uibutton(grid, 'Text', 'Button 1');
button2 = uibutton(grid, 'Text', 'Button 2');
editField = uieditfield(grid, 'Value', 'Enter Text');
text = uilabel(grid, 'Text', 'Some Text');
```

4. **Specifying Grid Layout for Each Component**

- button1.Layout = matlab.ui.layout.GridLayoutOptions('Row', 1, 'Column', 1);
  - This line specifies the position of `button1` within the grid.

    `matlab.ui.layout.GridLayoutOptions('Row', 1, 'Column', 1)` creates an object that defines the layout options for the button.

    `'Row', 1` specifies that the button should be placed in the first row of the grid.
    `'Column', 1` specifies that the button should be placed in the first column of the grid.

- The same logic applies to `button2`, `editField`, and `text`, placing them in the appropriate grid cells:

```
% Specify grid layout for each component
button1.Layout = matlab.ui.layout.GridLayoutOptions('Row',
1, 'Column', 1);
button2.Layout = matlab.ui.layout.GridLayoutOptions('Row',
1, 'Column', 2);
editField.Layout = matlab.ui.layout.GridLayoutOp-
tions('Row', 2, 'Column', 1);
text.Layout = matlab.ui.layout.GridLayoutOptions('Row', 2,
'Column', 2);
```

The entire function of the UIGrid App (described above) is given as follows:

```
function uigridApp
% Create the UI
app = uifigure('Name', 'UIGrid Example');

% Create a 2x2 grid layout
grid = uigridlayout(app, [2 2]);
% Create UI components
button1 = uibutton(grid, 'Text', 'Button 1');
button2 = uibutton(grid, 'Text', 'Button 2');
editField = uieditfield(grid, 'Value', 'Enter Text');
text = uilabel(grid, 'Text', 'Some Text');

% Specify grid layout for each component
button1.Layout = matlab.ui.layout.GridLayoutOptions('Row',
1, 'Column', 1);
button2.Layout = matlab.ui.layout.GridLayoutOptions('Row',
1, 'Column', 2);
editField.Layout = matlab.ui.layout.GridLayoutOp-
tions('Row', 2, 'Column', 1);
text.Layout = matlab.ui.layout.GridLayoutOptions('Row', 2,
'Column', 2);

end
```

To Use the App:

- **Save the code:** Save the code as an `.mlapp` file (e.g., **`UIGridApp.mlapp`**).
- **Open in App Designer:** Open the saved `.mlapp` file in MATLAB App Designer.
- **Run the App:** Click the "Run" button (green arrow) in the App Designer toolbar. This will launch the app (Fig. 3.9).

3.9 Conclusion

This chapter introduces the functionalities and usage of core UI components, such as Buttons, Radio Buttons, Sliders, and Checkboxes. It also covered the uigridlayout function in MATLAB App Designer, a powerful tool for organizing and positioning UI components within a grid structure. It explains how to use GridLayoutOptions to specify the exact location of each element within the grid, allowing for precise control over the layout. The chapter also presented the creation and use of fundamental UI elements, such as buttons (uibutton), edit fields (uieditfield), and labels (uilabel). It demonstrated how to integrate them effectively within the grid layout. By mastering these concepts, you can build more visually appealing and user-friendly MATLAB applications with well-structured and organized user interfaces.

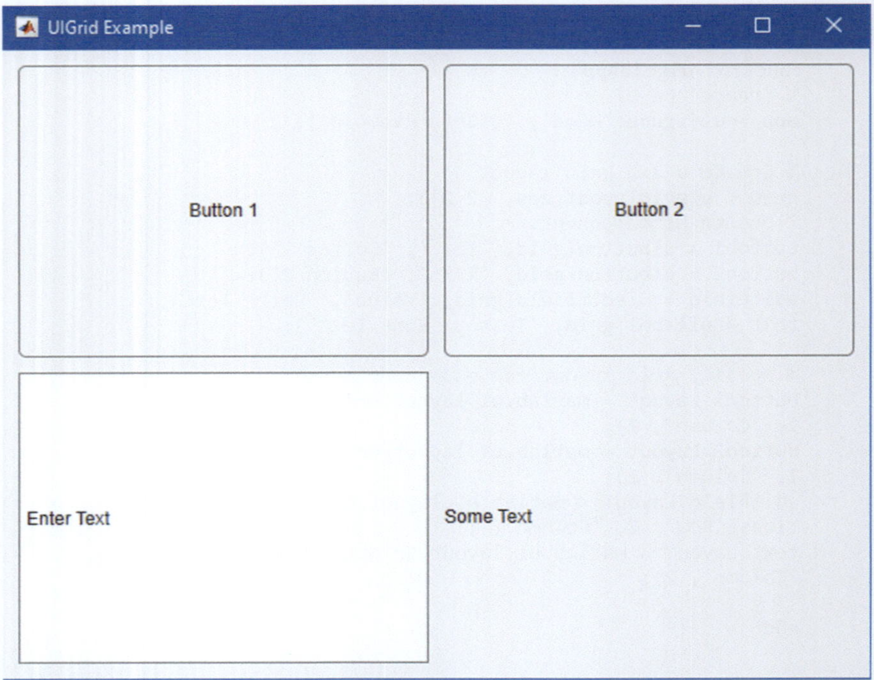

Fig. 3.9 An example of the MATLAB UIGrid App

References

1. Akan A, Chaparro LF. Signals and systems using MATLAB®. Academic Press; 2024.
2. Attaway DC. Matlab: a practical introduction to programming and problem solving. 6th ed. Butterworth-Heinemann; 2022.
3. C. Lopes P, L. Rangel R, Martha LF. An interactive user interface for a structural analysis software using computer graphics techniques in MATLAB. Computer Applications in Engineering Education. 2021;29(6):1505–25.
4. Hossain E. App Designer and Graphical User Interface in MATLAB. MATLAB and Simulink Crash Course for Engineers. Springer; 2022. p. 299–316.
5. Lent CS. Learning to program with MATLAB: Building GUI tools. John Wiley & Sons; 2022.
6. Parekh R. Fundamentals of image, audio, and video processing using Matlab®: with applications to pattern recognition. CRC Press; 2021.
7. Mathworks: MATLAB App Designer: Create desktop and web apps in MATLAB. https://au.mathworks.com/products/matlab/app-designer.html (2025). Accessed September 2025.
8. Valentine DT, Hahn BH. Essential MATLAB for engineers and scientists. Academic Press; 2022.

Chapter 4
Designing MATLAB App Layouts

4.1 Introduction

Designing a practical App layout is an essential step in creating intuitive and user-friendly MATLAB applications [1, 2]. A well-structured layout ensures users can easily interact with MATLAB Apps' features while maintaining visual clarity and logical flow. MATLAB App Designer provides a powerful drag-and-drop environment that permits you to arrange interface components such as buttons, sliders, images, plots, and text fields seamlessly [3–5].

In this chapter, we explore the key principles and tools for designing efficient MATLAB App layouts. Topics include Drag-and-Drop options, arranging components, using layout managers, aligning and resizing components for responsive design, and integrating User Interface (UI) controls with callback functions. By mastering these layout techniques, you can build professional, interactive applications that enhance usability and improve the overall user experience. Next, we explain how these UI components enable the creation of sophisticated and interactive MATLAB applications.

4.2 Drag and Drop

Drag-and-drop in MATLAB App Designer is a powerful, intuitive feature that significantly simplifies the creation of user interfaces. It lets you quickly and easily build visually appealing, functional Apps with minimal coding effort [2, 5].

Supplementary Information The online version contains supplementary material available at https://doi.org/10.1007/978-981-95-8797-1_4.

G. R. Naik, *MATLAB App Development*, https://doi.org/10.1007/978-981-95-8797-1_4

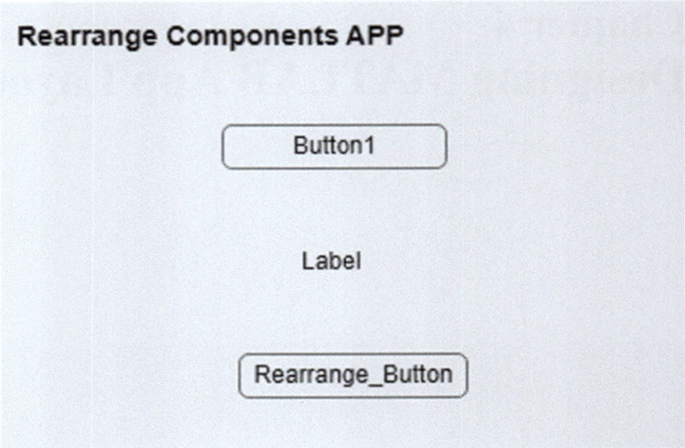

Fig. 4.1 Rearrange example components

Here is a simple example that demonstrates how to trigger component rearrangement with a button click, making the process more interactive. The entire procedure is explained as follows:

- **Open App Designer**: Start a new App Designer project.
- **Drag and Drop Components**: From the component library, drag and drop the desired components (Button1, label, and Rearrange button) onto the design canvas. In the right-side Inspector, under Component Browser → Text, name it to something meaningful like (Fig. 4.1)
  - Button1
  - Label
  - "Rearrange_Button"
- **Arrange Components**: Use the drag-and-drop interface to position and resize the components as needed.
- **Define Callback for Rearrange Button**: Right-click the "Rearrange" button → Callbacks → Add ButtonPushedFcn.

  App Designer will generate a function in Code View. Add the following position values for this example (The positions [150 80 100 30] are just an example; you can assign any desired [x y width height] values to rearrange.):

```
% Button pushed function: Rearrange_Button
function Rearrange_ButtonPushed(app, event)
% Example: move Button1 to a new position
% Swap positions of Button1 and Label
posBtn = app.Button1.Position;
posLbl = app.Label.Position;

app.Button1.Position = posLbl;
app.Label.Position = posBtn;
end
```

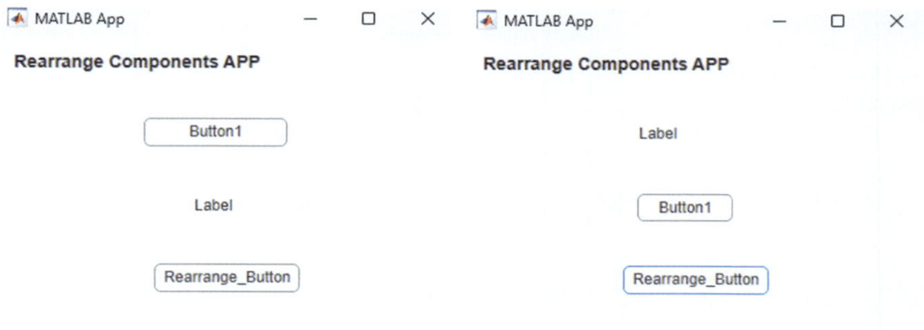

Fig. 4.2 Rearrange components example: **a** Before and **b** After

- **Save and Run the App**:
  - Save the completed App as "**Rearrange.mlapp**".
  - Your App will be launched in a new window.
  - When you press Rearrange_Button (Fig. 4.2a), Button1 and the Label move to the new positions (Fig. 4.2b).

4.3 Arranging Components

Resizing, moving, and aligning components using the Canvas toolbar is essential for MATLAB Apps. Here we illustrate two examples using the Canvas toolbar: (1) no coding is needed, and (2) programmatically arranging the tools in Apps and resetting them to the original positions [6, 7].

(1) **Arranging with the Canvas Toolbar**: The following simple example demonstrates how to resize, move, and align components using App Designer's Canvas toolbar. The entire process is explained as follows:

- **Open App Designer**: Start a new App Designer project.
- **Create a Panel**:
  - uipanel() creates a container within the figure to hold the components.
  - "Units" and "Position" control the size and position of the panel within the figure.
- **Create Example Components**: From the Component Library, drag the following onto the design canvas:
  - Label → set Text to "NameLabel."
  - Edit Field (Text) → for entering text, name as "Name Field."
  - Button → set Text to "RearrangeButton."

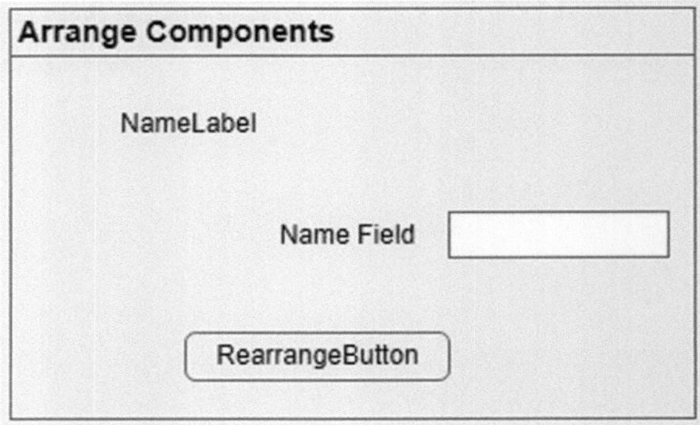

Fig. 4.3 Arranging components example—component placement

Place them randomly on the canvas. Now you have three UI components on your canvas (Fig. 4.3).

Use the Canvas Toolbar: On the Canvas toolbar (above the design area), you have tools to move, resize, and align components.

- **Resize**: Select the **Edit Field** and drag its edge to make it wider, so it can accept longer input.
- **Move**: Drag the "Submit" button below the Edit Field.
- **Align**
  - Select the Label and the Edit Field.
  - In the toolbar → click **Align > Align Left**.
  - Both components are now perfectly aligned vertically.
- **Distribute**
  - Select all three components (Label, Edit Field, and RearrangeButton).
  - In the toolbar → click **Evenly** > **Apply Vertically**.
  - Now the space between them is equal (Fig. 4.4).

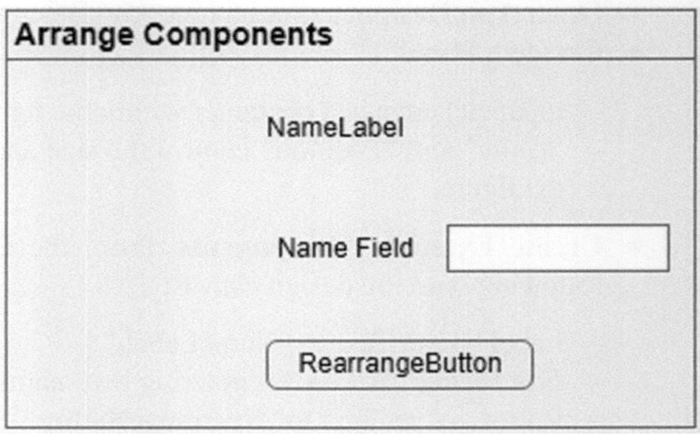

Fig. 4.4 Arranging components—components are vertically aligned

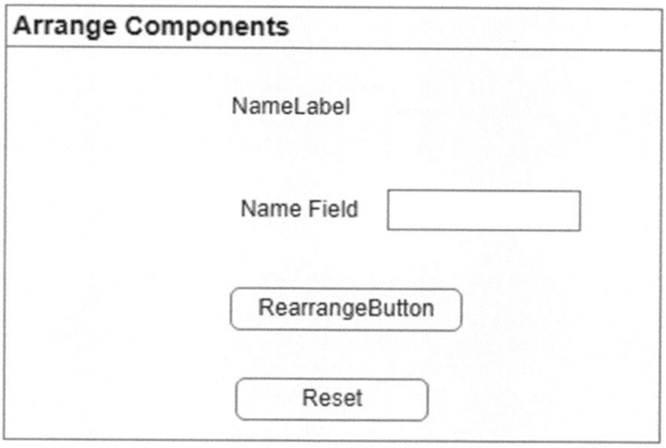

Fig. 4.5 Arranging components example—buttons and components are neatly arranged

(2) **Arranging with the Canvas Toolbar Programmatically with Reset Option**: The following simple example demonstrates how to resize, move, and align components using App Designer's Canvas toolbar programmatically. It also showcases MATLAB App's *startupFcn*, a special function that executes when the App starts up, but before the user interacts with the UI. The entire process is explained as follows:

- Components in the Component Browser (they are all neatly arranged—vertically):
  - NameLabel
  - NameField
  - RearrangeButton

Add a Button and rename it as "Reset" button (Fig. 4.5).

1. **Store the Original Positions When the App Starts**

Create a **startup function** from the **Code View → Callbacks → Add StartupFcn** menu and add the following code.

Note: MATLAB App Designer automatically generates the required *function* and *end* code skeletons. Please paste the remaining code into the appropriate sections within these generated functions.

```
function startupFcn(app)
            % Store original positions
                app.OriginalNameFieldPos = app.NameField.Po-
sition;
                app.OriginalNameLabelPos = app.NameLabel.Po-
sition;
                app.OriginalRearrangePos = app.RearrangeBut-
ton.Position;
                app.OriginalResetPos = app.ResetButton.Posi-
tion;
end
```

2. **Define the Property in a Properties Block**

Click the property, select a new private property, and add the following code:

```
properties (Access = private)
  OriginalNameFieldPos
  OriginalNameLabelPos
  OriginalRearrangePos
  OriginalResetPos
 end
```

3. **Add a Callback for Rearrange Button**

- Right-click RearrangeButton → *Callbacks* → *Add ButtonPushedFcn*
- **Insert the following Code in Code View**:

```
% Button pushed function: RearrangeButton
      function RearrangeButtonPushed(app, event)
            % Rearrange components dynamically
            % Move NameField down
            app.NameField.Position = [25 175,80,25];
            % Resize the Label
            app.NameLabel.Position = [90,125,105,25];
            % Move the Rearrange button to the right of
the text field
            app.RearrangeButton.Position =
[25,60,105,25];
            % Move the Reset button just below
            app.ResetButton.Position = [115,15,100,25];
        end
```

Add a Callback for Reset

- Right-click ResetButton → *Callbacks* → *Add ButtonPushedFcn*

- **Insert the following Code in Code View:**

```
% Button pushed function: ResetButton
      function ResetButtonPushed(app, event)
           % Reset to original positions
          app.NameField.Position = app.OriginalName-
  FieldPos;
          app.NameLabel.Position = app.OriginalName-
  LabelPos;
          app.RearrangeButton.Position = app.Origi-
  nalRearrangePos;
          app.ResetButton.Position = app.OriginalRese-
         Pos;
          end
```

4. **Save and Run the App:**

- Save the completed App as "**Arranging_Components.mlapp**".
- Your App will be launched in a new window.
- Press RearrangeButton.

Here, we observe the following: Before the click, the layout is neatly arranged (via the Canvas toolbar). However, **after clicking**, the components are rearranged programmatically (as shown in Fig. 4.6).

Here, positions are saved dynamically at startup and reused; hence, the Reset button always returns them to their original positions.

4.4 Layout Managers

Layout managers in MATLAB App Designer automatically control the size and position of UI components, which ensures the interface adapts smoothly to window resizing and different screen resolutions [6–8]. Here, we demonstrate how layout managers can be used for various applications in MATLAB. The following example shows the usage of the MATLAB App for split panels (splitting the panel as rows or columns).

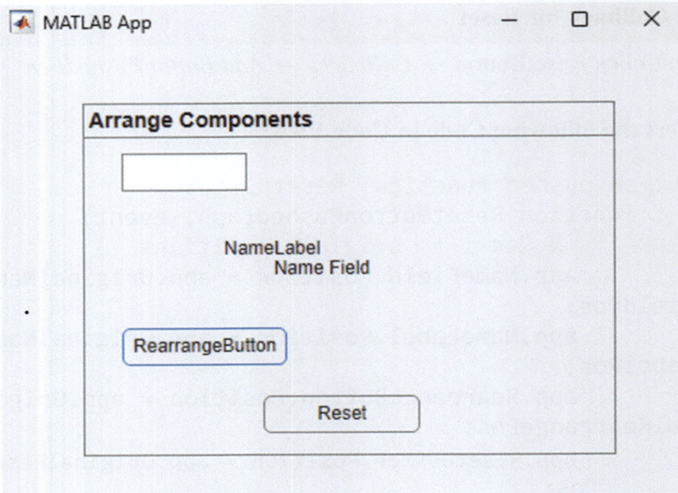

Fig. 4.6 Components changed after rearrangements

__**Split Panels Example**__: Here, we create a Split panel to display data on one side and a plot on the other. The entire process is explained as follows:

- **Create Figure**:
  - The figure creates the main figure window:

```
% Create the figure
fig = figure('Name', 'Split Panel App', 'Position',
[100 100 800 400]);
```

- **Calculate Panel Widths**:
  - Define the desired widths for the left and right panels:

```
% Calculate panel widths
panelWidth1 = 0.3;
panelWidth2 = 0.7;
```

- **Create Left Panel**:
  - uipanel creates a panel for the data display.
  - uicontrol creates an editable text box within the left panel:

```
% Create the left panel for data display
leftPanelPos = [0 0 panelWidth1 1];
leftPanel = uipanel('Parent', fig, 'Units', 'normal-
ized', 'Position', leftPanelPos);
dataDisplay = uicontrol('Parent', leftPanel, 'Style',
'edit', ...
    'String', 'Data will be displayed here', 'Units',
'normalized', 'Position', [0 0 1 1]);
```

- **Create Right Panel:**
  - uipanel creates a panel for the plot.
  - axes create the plotting area within the right panel:

```
% Create the right panel for plotting
rightPanelPos = [panelWidth1 0 panelWidth2 1];
rightPanel = uipanel('Parent', fig, 'Units', 'normal-
ized', 'Position', rightPanelPos);
plotAxes = axes('Parent', rightPanel, 'Units', 'nor-
malized', 'Position', [0 0 1 1]);
```

- **Position Panels**:
  - The Position property of each uipanel is set manually using the calculated widths to achieve the desired split.
- **Sample Data and Plotting**:
  - This section demonstrates how to plot sample data within the plotAxes.
  - Replace this with your actual data and plotting code:

```
% Sample data and plotting (if needed, replace with
your own data)
x = 1:0.01:100;
y = sin(x);
plot(plotAxes, x, y);
```

- **Update Data Display (Example)**:
  - set(dataDisplay, ‘String’, newData) shows how to update the text in the data display:

```
% Updating data display
newData = 'New data has been loaded.';
set(dataDisplay, 'String', newData);
```

The whole function of the "**splitPanelApp**" is given as follows:

```
function splitPanelApp
% Create the figure
fig = figure('Name', 'Split Panel App', 'Position', [100 100
800 400]);

% Calculate panel widths
panelWidth1 = 0.3;
panelWidth2 = 0.7;

% Create the left panel for data display
leftPanelPos = [0 0 panelWidth1 1];
leftPanel = uipanel('Parent', fig, 'Units', 'normalized',
'Position', leftPanelPos);
dataDisplay = uicontrol('Parent', leftPanel, 'Style',
'edit', ...
    'String', 'Data will be displayed here', 'Units', 'nor-
malized', 'Position', [0 0 1 1]);

% Create the right panel for plotting
rightPanelPos = [panelWidth1 0 panelWidth2 1];
rightPanel = uipanel('Parent', fig, 'Units', 'normalized',
'Position', rightPanelPos);


plotAxes = axes('Parent', rightPanel, 'Units', 'normalized',
'Position', [0 0 1 1]);

% Sample data and plotting (if needed, replace with your own
data)
x = 1:0.01:100;
y = sin(x);
plot(plotAxes, x, y);

% Example of updating data display
newData = 'New data has been loaded.';
set(dataDisplay, 'String', newData);

end
```

This will create a figure with a split panel (see Fig. 4.7). The left panel will display data, and the right panel will contain a plot. You can modify the data display, plotting code, and panel widths to suit your specific needs.

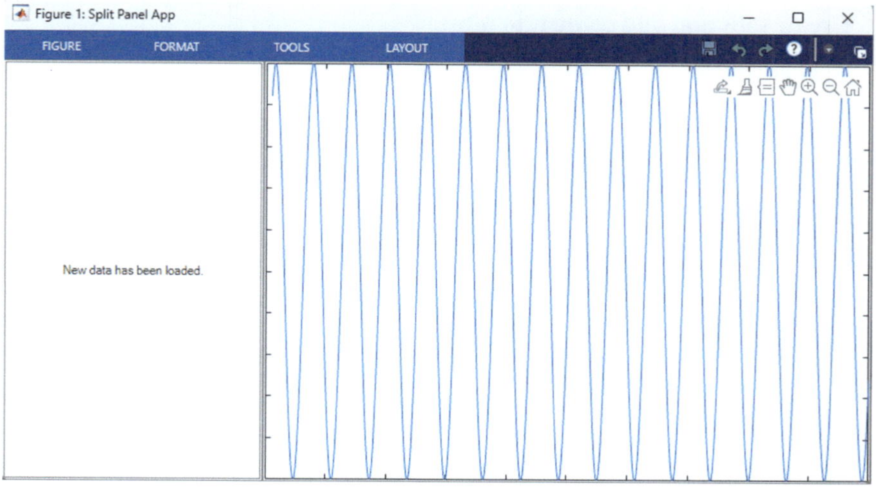

Fig. 4.7 Split panel example

Multiple Tabs Example: This example demonstrates how to organize a simple MATLAB App into multiple tabs. The process is explained in detail as follows:

- **Create a figure with multiple tabs**:
  - uifigure creates a new figure window with the specified title and position:

```
% Create a figure with multiple tabs
fig = uifigure('Name', 'Organized App', 'Position',
[100 100 600 400]);
```

- **Create a tab group**:
  - uitabgroup creates a container within the figure to hold multiple tabs:

```
% Create a tab group
tabGroup = uitabgroup(fig, 'Position', [0 0 600
400]);
```

- **Create tabs and add content to tabs**:
  - uitab creates individual tabs within the tab group. Each tab has a title that appears on the tab bar.
  - Within each tab, you can add any UI components using their respective creation functions:

    uibutton: Creates a button.
    uieditfield: Creates an editable text field.
    uiaxes: Creates axes for plotting:

```
% Create the first tab
tab1 = uitab(tabGroup, 'Title', 'Tab 1');

% Add content to the first tab (example: a button)
button1 = uibutton(tab1, 'Text', 'Button 1', 'Posi-
tion', [200 150 100 50]);

% Create the second tab
tab2 = uitab(tabGroup, 'Title', 'Tab 2');

% Add content to the second tab (example: an edit
field)
editField2 = uieditfield(tab2, 'Text', 'Position',
[200 150 200 30]);

% Create the third tab
tab3 = uitab(tabGroup, 'Title', 'Tab 3');

% Add content to the third tab (example: a plot)
axes3 = uiaxes(tab3, 'Position', [100 100 400 200]);
plot(axes3, 1:10, rand(1,10));
```

The whole function of the "**OrganizeAppTabs**" is given as follows:

```
function OrganizeAppTabs
% Create a figure with multiple tabs
fig = uifigure('Name', 'Organized App', 'Position',
[100 100 600 400]);

% Create a tab group
tabGroup = uitabgroup(fig, 'Position', [0 0 600
400]);

% Create the first tab
tab1 = uitab(tabGroup, 'Title', 'Tab 1');

% Add content to the first tab (example: a button)
button1 = uibutton(tab1, 'Text', 'Button 1', 'Posi-
tion', [200 150 100 50]);

% Create the second tab
tab2 = uitab(tabGroup, 'Title', 'Tab 2');

% Add content to the second tab (example: an edit
field)
editField2 = uieditfield(tab2, 'Text', 'Position',
[200 150 200 30]);

% Create the third tab
tab3 = uitab(tabGroup, 'Title', 'Tab 3');

% Add content to the third tab (example: a plot)
axes3 = uiaxes(tab3, 'Position', [100 100 400 200]);
plot(axes3, 1:10, rand(1,10));

end
```

- **To run this App:**
  - Save the code as an M-file (e.g., OrganizeAppTabs.m).
  - Run the script in MATLAB.

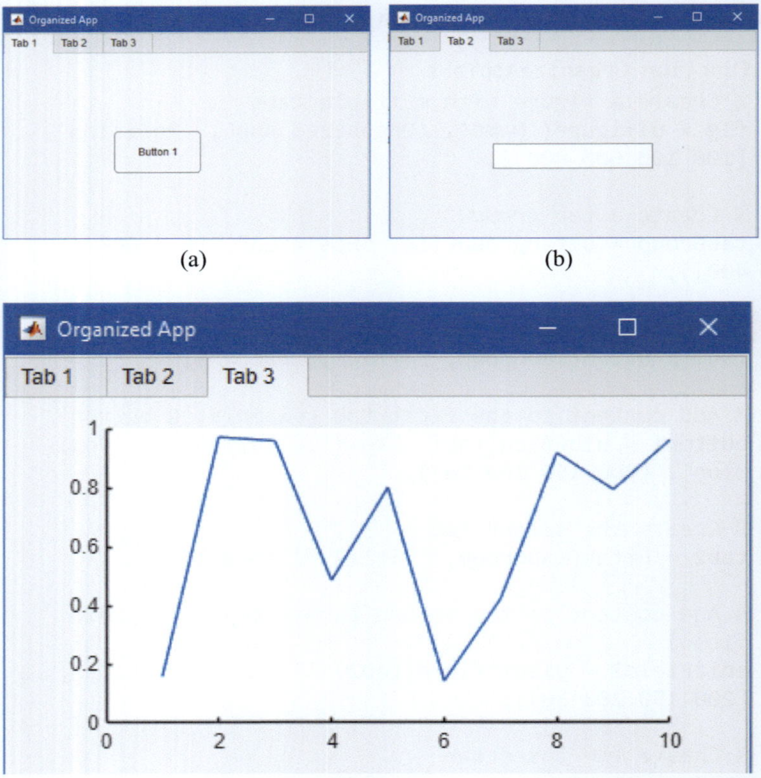

(a) (b)

(c)

Fig. 4.8 Organize App Example: **a** Tab1 shows a "Button," **b** Tab2 shows an "Edited field," and (3) Tab3 shows a "Plot"

– This will create a figure with three tabs, each containing a simple example of UI components (Fig. 4.8).

You can easily modify this code by

- Adding more tabs.
- Adding different UI components to each tab (e.g., sliders, dropdowns, images).
- Designing more complex layouts and interactions within each tab.

4.5 Responsive Design Considerations

Responsive design considerations are essential for effective MATLAB App design. This method provides a solid foundation for creating MATLAB Apps that adapt well to different screen sizes and resolutions, including PCs, laptops, tablets, and mobile devices [6–8]. Here, we demonstrate a simple MATLAB App that adapts well to different screen sizes and resolutions. The whole process is explained as follows:

- **Create Figure, add Normalized Units and UI components**: Using `'Units','normalized'` for the figure and grid ensures that component positions and sizes are relative to the figure's dimensions. This enables the App to automatically adjust to various screen sizes and resolutions:

```
% Create a figure with a flexible layout
fig = uifigure('Name', 'Responsive App', 'Units',
'normalized');

% Create a grid layout for flexible component ar-
rangement
grid = uigridlayout(fig, [2 2]); % 2 rows, 2 columns

% Create UI components
button1 = uibutton(grid, 'Text', 'Button 1', 'Hori-
zontalAlignment', 'center');
button2 = uibutton(grid, 'Text', 'Button 2', 'Hori-
zontalAlignment', 'center');
editField = uieditfield(grid, 'Text', 'Horizonta-
lAlignment', 'center');
axes1 = uiaxes(grid);
plot(axes1, rand(10,1));
```

- **Flexible Grid Layout**: The `uigridlayout` provides a flexible framework to arrange components. You can easily adjust the number of rows and columns and control the sizing behavior of each row and column (e.g., `'fit'`, `'auto'`, or fixed pixel values):

```
% Set grid properties for flexible layout
grid.RowHeight = {'2x', '8x'}; % Adjust row heights
as needed
grid.ColumnWidth = {'fit', 'fit'}; % Let columns ad-
just to content
```

- **Resize Function**: The `ResizeFcn` callback is triggered whenever the figure is resized. This allows you to dynamically adjust component sizes, positions, and other properties to maintain a good user experience across different screen sizes. The `resizeComponents` function demonstrates a simple way to adjust component properties based on the figure's dimensions. You can customize this function to handle more complex resizing scenarios:

```
function resizeComponents(fig, grid)
% Get figure dimensions
figWidth = fig.Position(3);
figHeight = fig.Position(4);
% Adjust component sizes and positions (example)
grid.RowHeight = {0.2*figHeight, 0.8*figHeight};
end
```

The whole function of the "**responsiveApp**" is given as follows:

```
function responsiveApp
% Create a figure with a flexible layout
fig = uifigure('Name', 'Responsive App', 'Units',
'normalized');

% Create a grid layout for flexible component ar-
rangement
grid = uigridlayout(fig, [2 2]); % 2 rows, 2 columns

% Create UI components
button1 = uibutton(grid, 'Text', 'Button 1', 'Hori-
zontalAlignment', 'center');
button2 = uibutton(grid, 'Text', 'Button 2', 'Hori-
zontalAlignment', 'center');
editField = uieditfield(grid, 'Text', 'Horizonta-
lAlignment', 'center');
axes1 = uiaxes(grid);
plot(axes1, rand(10,1));

% Set grid properties for flexible layout
grid.RowHeight = {'2x', '8x'}; % Adjust row heights
as needed
grid.ColumnWidth = {'fit', 'fit'}; % Let columns ad-
just to content

% Set figure resize behavior
fig.ResizeFcn = @(~,~) resizeComponents(fig, grid);

end
```

```
function resizeComponents(fig, grid)

% Get figure dimensions
figWidth = fig.Position(3);
figHeight = fig.Position(4);

% Adjust component sizes and positions (example)
grid.RowHeight = {0.2*figHeight, 0.8*figHeight};
end
```

To use this code:

- Save the code as an M-file (e.g., `responsiveApp.m`).
- Run the script in MATLAB.

This will create a simple App with a button, an edit field, and a plot (Fig. 4.9). Resize the figure window to see how the components automatically adjust their positions and sizes.

Fig. 4.9 Organize App Example where the components are automatically adjusted based on their sizes and positions

Further Enhancements:

- Consider using uipanel: For more complex layouts, consider using uipanel objects within the grid to group related components and control their resizing behavior as a unit.
- Implement more sophisticated resizing logic: The **resizeComponents** function can be enhanced to meet the specific needs of your App. For example, you can use conditional statements to adjust components differently based on the figure's aspect ratio or size.
- Test on different screen sizes and resolutions: Thoroughly test your App on various devices and screen resolutions to ensure it behaves as expected.

This approach provides a solid foundation for creating responsive MATLAB Apps that adapt well to different screen sizes and resolutions. Remember to adjust the code and layout to fit the specific requirements of your application.

4.6 Conclusion

This chapter provides a comprehensive guide to designing effective layouts in MATLAB's App Designer, a crucial skill for developing sophisticated, interactive applications. We explored fundamental techniques, starting with the intuitive Drag-and-Drop method, which allows quick placement of UI components. Building on this, we delved into various strategies for Arranging Components, emphasizing precision and organization to create visually appealing, user-friendly interfaces.

A significant portion of this chapter was dedicated to understanding and utilizing Layout Managers. We demonstrated how these powerful tools automate component positioning and sizing, ensuring consistency and adaptability across different screen sizes and resolutions. Finally, we discussed key Responsive Design Considerations, highlighting how to create applications that maintain functionality and aesthetic appeal across user display environments. By mastering the principles and techniques outlined in this chapter, developers can create robust, visually appealing, and highly interactive MATLAB applications that deliver a superior user experience.

References

1. Valentine DT, Hahn BH. Essential MATLAB for engineers and scientists. Academic Press; 2022.
2. Hossain E. App Designer and Graphical User Interface in MATLAB. MATLAB and Simulink Crash Course for Engineers. Springer; 2022. p. 299–316.
3. Lent CS. Learning to program with MATLAB: Building GUI tools. John Wiley & Sons; 2022.
4. Mathworks: MATLAB App Designer: Create desktop and web apps in MATLAB. https://au.mathworks.com/products/matlab/app-designer.html (2025). Accessed September 2025.

5. Thabit AA, Alzidi E, Mohammed MA, Palchyk P. Development of a System Based on a Graphical User Interface Using MATLAB Tools. 2024 35th Conference of Open Innovations Association (FRUCT): IEEE; 2024. p. 744–54.
6. Darweesh M. Teaching MATLAB Programming by Utilizing Image Processing and Graphical User Interface Basics. Interactive Mobile Communication, Technologies and Learning. Springer; 2021. p. 824–32.
7. Attaway DC. Matlab: a practical introduction to programming and problem solving. 6th ed. Butterworth-Heinemann; 2022.
8. Akan A, Chaparro LF. Signals and systems using MATLAB®. Academic Press; 2024.

Chapter 5
Writing MATLAB App Code

5.1 Introduction

Writing App Code in MATLAB App Designer is where you bring your graphical user interface to life [1–3]. While Design View lets you arrange components visually, Code View is where you define the behavior and logic that make your App interactive and functional. This involves writing MATLAB code within specific functions and callbacks to respond to user actions, process data, update the User Interface (UI), and implement the core functionality of your application [4, 5].

This chapter explains how to write callbacks, perform simple calculations, and manage App properties while building MATLAB Apps. Below, we explain how these components are used to create sophisticated, interactive MATLAB applications.

5.2 Callbacks

A **callback** in MATLAB App Designer is a function or piece of code that automatically executes in response to a specific **event** occurring in your App. It is how you define the behavior and interactivity of your UI components [4, 5].

For example, when a user clicks a button, changes a slider's value, or selects an item from a dropdown list, an event is triggered, etc. The callback associated with that event and component is then executed, allowing your App to perform actions such as updating a plot, displaying a message, and processing data. You define these callbacks within the App Designer's integrated editor.

Supplementary Information The online version contains supplementary material available at https://doi.org/10.1007/978-981-95-8797-1_5.

G. R. Naik, *MATLAB App Development*, https://doi.org/10.1007/978-981-95-8797-1_5

Here, we create a simple App, a callback function for a button ("**Calculate**") that performs a calculation and updates the output. Steps are given as follows:

1. **Create the UI with the Following Components:**

- `uifigure`: creates the main figure window for the App.
- `uieditfield`: creates input fields for the two numbers.
- `uilabel`: creates labels for input and output numbers.
- `uidropdown`: creates a dropdown list for selecting the operators (+, −, *, and /).
- `uibutton`: creates the "Calculate" button with the `ButtonPushedFcn` property set to the `calculateCallback` function:

```
% Create the UIFigure and components
fig = uifigure('Name', 'Simple Calculator');

input1Edit = uieditfield(fig, 'numeric', 'Posi-
tion', [50 150 100 22], ...
    'Value', 0);
input1Label = uilabel(fig, 'Position', [50 170 100
15], 'Text', 'Input 1:');

input2Edit = uieditfield(fig, 'numeric', 'Posi-
tion', [50 100 100 22], ...
    'Value', 0);
input2Label = uilabel(fig, 'Position', [50 120 100
15], 'Text', 'Input 2:');

operatorDropdown = uidropdown(fig, 'Position', [50
50 100 22], ...
    'Items', {'+', '-', '*', '/'}, 'Value', '+');
% Corrected Value
operatorLabel = uilabel(fig, 'Position', [50 70
100 15], 'Text', 'Operator:');

calculateButton = uibutton(fig, 'push', 'Posi-
tion', [50 20 100 22], ...
    'Text', 'Calculate', 'ButtonPushedFcn',
@(src,event) calculateCallback(src, event, fig,
input1Edit, input2Edit, operatorDropdown));

outputText = uilabel(fig, 'Position', [200 100 150
22], 'Text', 'Result:');
outputValue = uilabel(fig, 'Position', [200 50 150
22]);
```

2. **Define the Callback Function (Calculate)**:
   - `calculateCallback`: This function is executed when the "Calculate" button is pressed.
   - It retrieves the values from the input fields and the selected operator.
   - It performs the calculation based on the chosen operator (+, −, *, and /).
   - It updates the outputValue label with the calculated result:

```
% Callback function for the Calculate button
function calculateCallback(~, ~, fig, input1Edit,
input2Edit, operatorDropdown)
    input1 = input1Edit.Value;
    input2 = input2Edit.Value;
    operator = operatorDropdown.Value;

    switch operator
        case '+'
            result = input1 + input2;
        case '-'
            result = input1 - input2;
        case '*'
            result = input1 * input2;
        case '/'
            if input2 == 0
                result = 'Error: Division by
zero';
            else
                result = input1 / input2;
            end
    end
    outputValue.Text = num2str(result);
end
```

The whole function of the "**MyCalculatorApp**" is given as follows:

Note: MATLAB App Designer automatically generates the required *function* and *end* code skeletons. Please paste the remaining code into the appropriate sections within these generated functions.

```
function MyCalculatorApp

% Create the UIFigure and components
fig = uifigure('Name', 'Simple Calculator');

input1Edit = uieditfield(fig, 'numeric', 'Posi-
tion', [50 150 100 22], ...
    'Value', 0);
input1Label = uilabel(fig, 'Position', [50 170 100
15], 'Text', 'Input 1:');
input2Edit = uieditfield(fig, 'numeric', 'Posi-
tion', [50 100 100 22], ...
    'Value', 0);
input2Label = uilabel(fig, 'Position', [50 120 100
15], 'Text', 'Input 2:');

operatorDropdown = uidropdown(fig, 'Position', [50
50 100 22], ...
    'Items', {'+', '-', '*', '/'}, 'Value', '+');
% Corrected Value
operatorLabel = uilabel(fig, 'Position', [50 70
100 15], 'Text', 'Operator:');

calculateButton = uibutton(fig, 'push', 'Posi-
tion', [50 20 100 22], ...
    'Text', 'Calculate', 'ButtonPushedFcn',
@(src,event) calculateCallback(src, event, fig,
input1Edit, input2Edit, operatorDropdown));

outputText = uilabel(fig, 'Position', [200 100 150
22], 'Text', 'Result:');
outputValue = uilabel(fig, 'Position', [200 50 150
22]);

% Callback function for the Calculate button
function calculateCallback(~, ~, fig, input1Edit,
input2Edit, operatorDropdown)
    input1 = input1Edit.Value;
    input2 = input2Edit.Value;
    operator = operatorDropdown.Value;
```

```
            switch operator
                case '+'
                    result = input1 + input2;
                case '-'
                    result = input1 - input2;
                case '*'
                    result = input1 * input2;
                case '/'
                    if input2 == 0
                        result = 'Error: Division by
        zero';
                    else
                        result = input1 / input2;
                    end
            end
            outputValue.Text = num2str(result);
        end
        end
```

3. **Run the App:**
   - Save the code as an `.m file` (e.g., **`MyCalculatorApp.m`**).
   - Run the script in MATLAB (either from the MATLAB command window or using the Run function—green arrow).

This will create a simple calculator App with a user-friendly interface and a functional "**Calculate**" button (Fig. 5.1).

5.3 App Properties

In MATLAB App Designer, **app properties** are variables that belong to your App and are used to store data, manage the state of your App, and share information between different callbacks and functions within your application. They are fundamental for making your App interactive and functional [4–6].

Here we outline key components:

- **Sharing Data**: Properties are the primary mechanism for sharing data throughout your App. If you have a value calculated in one callback (e.g., from a button click) that needs to be used by another callback (e.g., to update a plot), you store that value in an App property.
- **Accessing UI Components**: All the UI components you drag onto the canvas (buttons, plots, text fields, etc.) are themselves properties of your App. This allows you to access and modify their characteristics programmatically (such as Value, Text, Color, and Visible state) in

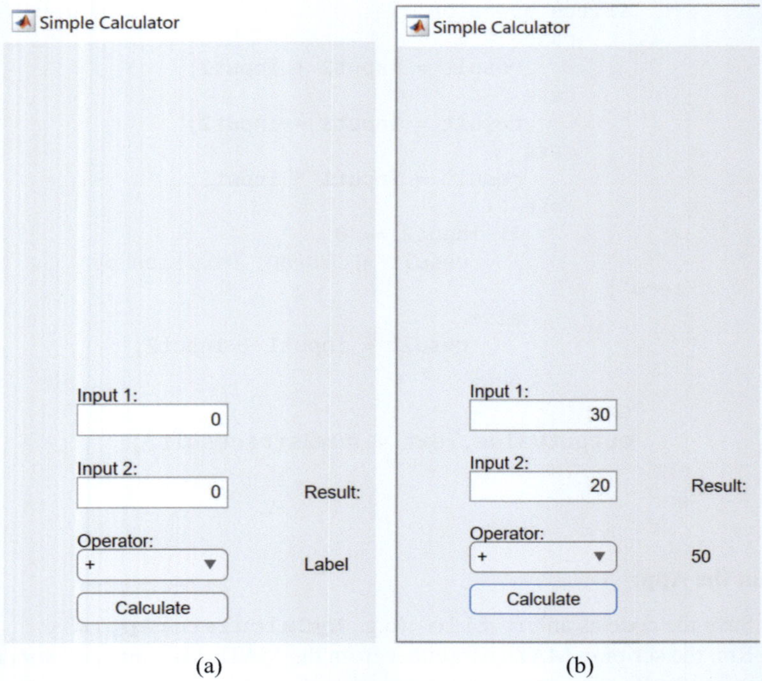

Fig. 5.1 Simple Calculator App: **a** Simple calculator layout and **b** Simple arithmetic operation demonstration

your code using the `app.ComponentName.PropertyName` syntax (e.g., `app.MyButton.Text = 'Clicked!';`).

- **Defining Custom Properties**: In addition to the built-in properties for UI components, you can define custom properties to store data specific to your App's logic. These can be `public property,` accessible from outside the App and useful for custom components, or private property, accessible only within the App and *ideal for internal data.*
- **Initialization**: You can initialize property values directly when you define them or within the app's `startupFcn` callback, which runs when the App is launched.

Below, we discuss some of the App properties, along with examples that are frequently used during MATLAB App design.

Accessing Component Properties:

- **Read and modify properties of UI components within the code:**

Here, we create an App that reads the value entered in a text box and changes the color of a plot. We include a `try-catch` block to handle invalid color inputs, making it more robust. The code displays an error message in the text box if an invalid color is entered, providing feedback to the user.

- **Create the following UI**:
  - figure: creates the main figure window.
  - plot: creates a simple sine wave plot with an initial blue color.
  - uicontrol: creates the text box for color input and the "Update Plot" button:

```
% Create the figure
f = figure('Name', 'Color Plot App', 'Position', [100
100 400 300]);

% Create the plot
x = 0:0.1:10;
y = sin(x);
plot(x, y, 'b-'); % Initial plot in blue
% Create the text box for color input
color_input = uicontrol('Style', 'edit', 'String', 'b',
'Position', [50 250 200 25]);
% Create the update button
update_button = uicontrol('Style', 'pushbutton',
'String', 'Update Plot', 'Position', [120 200 100 25],
...
                              'Callback', @update_plot);
```

- **update_plot Callback:**
  - get(color_input, 'String'): Retrieves the color string entered by the user.
  - set(findobj(gca, 'Type', 'line'), 'Color', color_str): Attempts to update the color of the plotted line using the input string.
  - try-catch block: Handles potential errors, such as invalid color strings. If an error occurs, it displays an error message in the text box:

```
% Callback function to update the plot
    function update_plot(~, ~)
        % Get the color input from the text box
        color_str = get(color_input, 'String');

        % Attempt to update the plot color
        try
            set(findobj(gca, 'Type', 'line'), 'Color',
color_str);
        catch
            % Handle invalid color input
            set(color_input, 'String', 'Invalid color');
        end
    end
```

The whole function of the "**color_plot_app**" is given as follows:

```
function color_plot_app
% Create the figure
f = figure('Name', 'Color Plot App', 'Position', [100
100 400 300]);

% Create the plot
x = 0:0.1:10;
y = sin(x);
plot(x, y, 'b-'); % Initial plot in blue
% Create the text box for color input
color_input = uicontrol('Style', 'edit', 'String', 'b',
'Position', [50 250 200 25]);

% Create the update button
update_button = uicontrol('Style', 'pushbutton',
'String', 'Update Plot', 'Position', [120 200 100 25],
...
                               'Callback', @update_plot);
% Callback function to update the plot
    function update_plot(~, ~)
        % Get the color input from the text box
        color_str = get(color_input, 'String');

        % Attempt to update the plot color
        try
            set(findobj(gca, 'Type', 'line'), 'Color',
color_str);
        catch
            % Handle invalid color input
            set(color_input, 'String', 'Invalid color');
        end
    end
end
```

To Use the App, Do the Following:

- Save the code as an M-file (e.g., **`color_plot_app.m`**).
- Run the M-file in MATLAB.
- Enter a valid color string (e.g., "r" for red, "g" for green, "b" for blue, "k" for black, "m" for magenta, "c" for cyan, "y" for yellow, or RGB triplets like "[0.8 0.2 0.1]") in the text box.
- Click the "Update Plot" button. The plot will update with the specified color (Fig. 5.2).

Callbacks are highly useful for performing mathematical calculations. They also help us process data, perform computations, and generate results effectively. Here, we provide an example of implementing mathematical functions and performing data analysis. The App computes the Mean and Standard Deviation of numbers entered in the Text box. The process is described as follows:

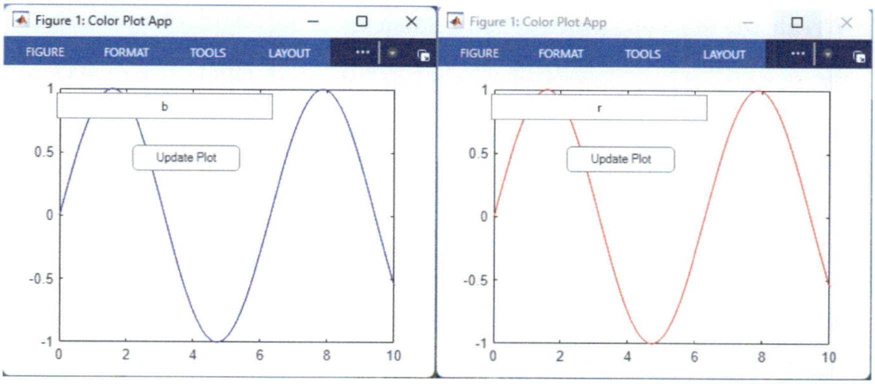

Fig. 5.2 Example of the Update plot function: **a** Updated plot using "blue" color and **b** updated plot using "red" color

1. **Create the following UI Components**:
   - `figure`: creates the main figure window.
   - `uicontrol`: creates the UI elements:
     - `edit`: creates a text box for user input.
     - `pushbutton`: creates a button to trigger the data processing.
     - `text`: creates a static text box to display the results:

```
f = figure('Name', 'Data Processing App', 'Posi-
tion', [100 100 400 300]);

% Create UI components
input_edit = uicontrol('Style', 'edit', 'String',
'', 'Position', [50 250 200 25]);
process_button = uicontrol('Style', 'pushbutton',
'String', 'Process', 'Position', [120 200 100 25],
...
                          'Callback', @pro-
cess_data);
result_text = uicontrol('Style', 'text', 'String',
'', 'Position', [50 150 200 25]);
```

2. **Create the following process data Callback**:

- `get(input_edit, 'String')`: retrieves the input string from the text box.
- `str2num(input_str)`: converts the input string into a numerical array.
- `mean(data)` and `std(data)`: calculates the mean and the standard deviation of the input data.
- `sprintf`: formats the results into a string for display.
- `set(result_text, 'String', result_str)`: updates the result text box with the calculated values.
- Error handling: checks for empty input and displays an error (invalid) message:

```
% Callback function to process data
  function process_data(~, ~)
        % Get input data (example: assume input is
a comma-separated list of numbers)
        input_str = get(input_edit, 'String');
        data = str2num(input_str);

        % Perform computations (example: calculate
mean and standard deviation)
        if ~isempty(data)
            mean_value = mean(data);
            std_dev = std(data);

            % Display results
            result_str = sprintf('Mean:
%.2f\nStandard Deviation: %.2f', mean_value,
std_dev);
            set(result_text, 'String', re-
sult_str);
        else
            set(result_text, 'String', 'Invalid
input.');
        end
    end
```

The whole function of the "**data_processing_app**" is given as follows:

```
function data_processing_app

% Create the figure
f = figure('Name', 'Data Processing App', 'Posi-
tion', [100 100 400 300]);
% Create UI components
input_edit = uicontrol('Style', 'edit', 'String',
'', 'Position', [50 250 200 25]);
process_button = uicontrol('Style', 'pushbutton',
'String', 'Process', 'Position', [120 200 100 25],
...
                                  'Callback', @pro-
cess_data);
result_text = uicontrol('Style', 'text', 'String',
'', 'Position', [50 150 200 25]);

% Callback function to process data
    function process_data(~, ~)
        % Get input data (example: assume input is
a comma-separated list of numbers)
        input_str = get(input_edit, 'String');
        data = str2num(input_str);

        % Perform computations (example: calculate
mean and standard deviation)
        if ~isempty(data)
            mean_value = mean(data);
            std_dev = std(data);

            % Display results
            result_str = sprintf('Mean:
%.2f\nStandard Deviation: %.2f', mean_value,
std_dev);
            set(result_text, 'String', re-
sult_str);
        else
            set(result_text, 'String', 'Invalid
input.');
        end
    end

end
```

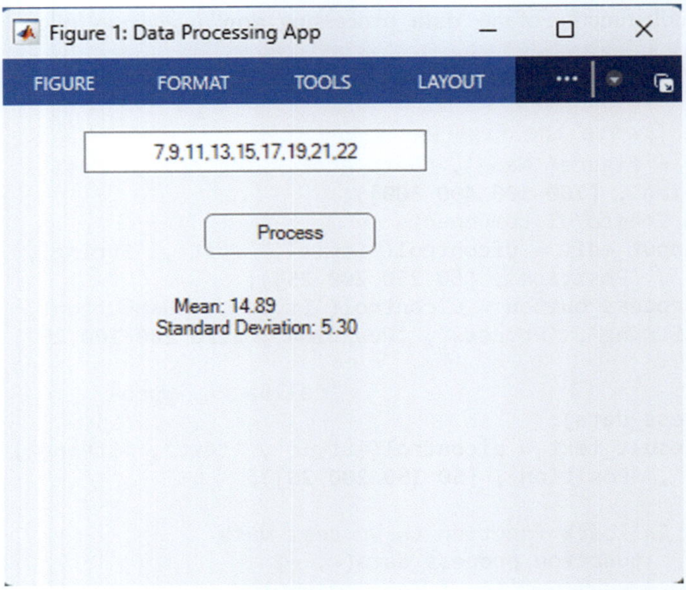

Fig. 5.3 Mean and standard deviation plot

To Use This App, Do the Following:

- Save the code as an M-file (e.g., **`data_processing_app.m`**).
- Run the M-file in MATLAB.
- Enter a comma-separated list of numbers in the text box (e.g., "7,9,11,13,15,17,19,21,22").
- Click the "Process" button.
- The calculated Mean and Standard Deviation will be displayed in the result text box (Fig. 5.3).

This App provides a basic framework for building data processing applications in MATLAB. You can customize and expand it to suit your specific needs and complexity.

5.4 Updating the UI

Updating the UI in MATLAB App Designer refers to programmatically changing the visual appearance or content of your App's components while the App is running. This is typically done within **callbacks** triggered by user actions or other events [7, 8].

Why Update the UI?

- **Provide Feedback**: Indicate to the user that an action has been performed (e.g., changing a button's text to "Processing…").
- **Display Results**: Show new data or calculation outcomes (e.g., updating a plot, populating a table, displaying text in a label).
- **Change State**: Enable or turn off components based on logic (e.g., disabling a "Start" button once a process has begun).
- **Reflect Data Changes**: Ensure the displayed information accurately reflects the underlying data.

Here, we provide an example of updating a plot when new data is loaded or a parameter is changed. The steps are given as follows:

1. **Create the following UI Components**:

   - figure: creates the main figure window.
   - plot: creates the initial plot with sample data:

```
% Create the figure
f = figure('Name', 'Dynamic Plot App', 'Position',
[100 100 400 300]);

% Create the plot
x = 0:0.1:10;
y = sin(x);
hPlot = plot(x, y, 'b-');
xlabel('Time');
ylabel('Signal');
title('Dynamic Plot');
```

2. **Create the following uicontrol elements:**

   - create a slider to control a plot parameter (in this case, amplitude).
   - create a button to load new data:

```
% Create the parameter slider
slider = uicontrol('Style', 'slider', 'Min', 1,
'Max', 5, 'Value', 1, 'Position', [50 250 200 25],
...
                  'Callback', @update_plot);

% Create the load data button
load_button = uicontrol('Style', 'pushbutton',
'String', 'Load Data', 'Position', [120 200 100 25],
...
                        'Callback', @load_data);
```

3. **Update the plot Callback:**

- Gets the current value of the slider.
- Updates the YData of the plot based on the slider value.
- set(hPlot, 'YData', y): Modifies the plot data without recreating the entire plot:

```
% Callback function to update the plot based on
slider value
    function update_plot(~, ~)
    % Get slider value
    slider_value = get(slider, 'Value');

    % Update the plot data (example: modify am-
plitude)
    y = slider_value * sin(x);
    set(hPlot, 'YData', y);
end
```

4. **Load data Callback:**

- Loads new data (example: random data).
- Updates both XData and YData of the plot with the new data:

```
% Callback function to load new data
    function load_data(~, ~)
        % Load new data (example: random data)
        x = 0:0.1:10;
        y = rand(size(x));

        % Update the plot data
        set(hPlot, 'XData', x, 'YData', y);
    end
```

The whole function of the "**dynamic_plot_app**" is given as follows:

```
function dynamic_plot_app
% Create the figure
f = figure('Name', 'Dynamic Plot App', 'Position',
[100 100 400 300]);

% Create the plot
x = 0:0.1:10;
y = sin(x);
hPlot = plot(x, y, 'b-');
xlabel('Time');
ylabel('Signal');
title('Dynamic Plot');

% Create the parameter slider
slider = uicontrol('Style', 'slider', 'Min', 1,
'Max', 5, 'Value', 1, 'Position', [50 250 200 25],
...
                    'Callback', @update_plot);


% Create the load data button
load_button = uicontrol('Style', 'pushbutton',
'String', 'Load Data', 'Position', [120 200 100 25],
...
                         'Callback', @load_data);

% Callback function to update the plot based on
slider value
    function update_plot(~, ~)
        % Get slider value
        slider_value = get(slider, 'Value');

        % Update the plot data (example: modify am-
plitude)
        y = slider_value * sin(x);
        set(hPlot, 'YData', y);
    end

% Callback function to load new data
    function load_data(~, ~)
        % Load new data (example: random data)
        x = 0:0.1:10;
        y = rand(size(x));

        % Update the plot data
        set(hPlot, 'XData', x, 'YData', y);
    end

end
```

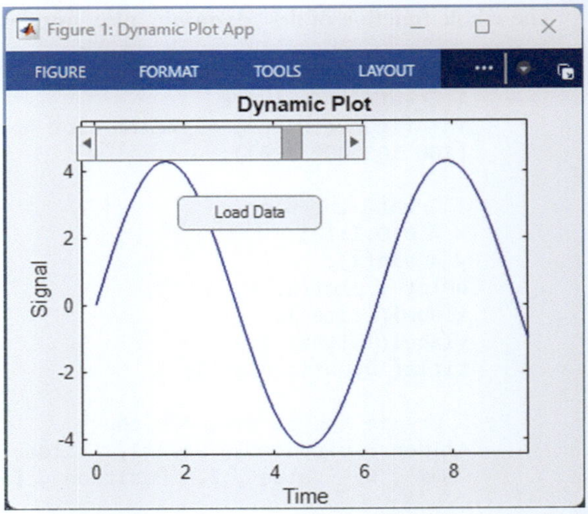

Fig. 5.4 Dynamic sine wave plot

To Use This App, Please Do the Following:

1. Save the code as an M-file (e.g., **dynamic_plot_app.m**).
2. Run the M-file in MATLAB.
3. Adjust the slider to see the plot update dynamically.
4. Click the "Load Data" button to load new data and update the plot (Fig. 5.4).

This App provides a basic framework for creating dynamic plots in MATLAB, which include

- Dynamic Updates: The set function efficiently updates only the necessary parts of the plot, resulting in smoother and faster updates.
- Parameter Control: The slider enables interactive control of a plot parameter, allowing for real-time exploration and visualization.
- Data Loading: The load_data callback demonstrates how to incorporate new data into the plot.

You can adapt and extend it to suit your specific data analysis and visualization needs, which include the following:

- Add more parameters to control (e.g., frequency, phase).
- Allow users to select different data loading options or import data from files.
- Implement more sophisticated data visualization techniques (e.g., zooming, panning).
- Create a more user-friendly interface with better visual cues.

5.5 Conclusion

In conclusion, this chapter has set the essential foundation for building robust, interactive MATLAB Apps. We have explored how callbacks serve as the fundamental mechanism for responding to user actions, transforming static interfaces into dynamic experiences. Understanding App properties has proven vital for managing application state, storing data, and allowing different parts of your App to communicate seamlessly. Finally, we have demonstrated the importance of effectively updating the UI to provide real-time visual feedback and ensure a responsive user experience.

By mastering these core components, callbacks for interaction, properties for data management, and UI updates for responsiveness, you are now prepared to address a wide range of application development challenges. These principles form the basis for creating more sophisticated and powerful MATLAB Apps that integrate advanced features and deliver highly intuitive user interfaces.

References

1. Valentine DT, Hahn BH. Essential MATLAB for engineers and scientists. Academic Press; 2022.
2. Hossain E. App Designer and Graphical User Interface in MATLAB. MATLAB and Simulink Crash Course for Engineers. Springer; 2022. p. 299–316.
3. Lent CS. Learning to program with MATLAB: Building GUI tools. John Wiley & Sons; 2022.
4. Mathworks: MATLAB App Designer: Create desktop and web apps in MATLAB. https://au.mathworks.com/products/matlab/app-designer.html (2025). Accessed September 2025.
5. Thabit AA, Alzidi E, Mohammed MA, Palchyk P. Development of a System Based on a Graphical User Interface Using MATLAB Tools. 2024 35th Conference of Open Innovations Association (FRUCT): IEEE; 2024. p. 744–54.
6. Akan A, Chaparro LF. Signals and systems using MATLAB®. Academic Press; 2024.
7. Darweesh M. Teaching MATLAB Programming by Utilizing Image Processing and Graphical User Interface Basics. Interactive Mobile Communication, Technologies and Learning. Springer; 2021. p. 824–32.
8. Attaway DC. Matlab: a practical introduction to programming and problem solving. 6th ed. Butterworth-Heinemann; 2022.

Part III
App Logic and Behavior

Chapter 6
Handling User Input

6.1 Introduction

In MATLAB Apps, user input is usually collected via interactive components such as buttons, edit text boxes, and drop-down menus [1, 2]. These inputs are then processed by callback functions that are triggered when the user interacts with the component (e.g., pressing a button). To ensure robustness of MATLAB Apps, user input must be validated before performing calculations or updating results. For instance, checking for numeric values or preventing division by zero [2–4].

This approach makes MATLAB Apps more interactive, user-friendly, and error resilient. This chapter explains how to handle user inputs in MATLAB Apps. Core components are presented with several examples.

6.2 Data Validation and Input Sanitization

Data validation and input sanitization are essential for creating efficient MATLAB Apps [5, 6]. They involve the following crucial steps:

- Check user input for validity (e.g., correct data types, within expected ranges).
- Sanitize input to prevent errors or security vulnerabilities.

Supplementary Information The online version contains supplementary material available at https://doi.org/10.1007/978-981-95-8797-1_6.

G. R. Naik, *MATLAB App Development*, https://doi.org/10.1007/978-981-95-8797-1_6

Here, we create an example and check if a text box input is a valid number before performing calculations. The steps are outlined as follows:

1. **Create the following UI Components**:
   - **figure**: Use figure UI to create the main window of the App:

```
% Create the UI components
f = figure('Name','Number Checker App', 'Position',
[100 100 300 150]);
```

   - **uicontrol**: Creates the following UI elements:
     - **edit**: create the input text boxes for input and results display.
     - **pushbutton**: create the button to trigger the calculation:

```
% Text box for input
input_box = uicon-
trol('Style','edit','String','','Position',[50
100 200 25]);

% Button to trigger calculation
calculate_button = uicontrol('Style','pushbut-
ton','String','Calculate','Position',[90 50 100
25],...
                                        'Callback',@cal-
culate_callback);

% Text box to display result
result_box = uicon-
trol('Style','edit','String','','Position',[50
20 200 25],'Enable','inactive');
```

2. **Calculate the callback Function:**
   - This function is executed when the “**Calculate**” button is clicked.
   - get(input_box, ‘String’): Retrieves the text entered by the user.

- str2double(input_str): Attempts to convert the input string to a number.
- isnumeric(…): Checks if the conversion was successful.
  - If successful, perform the calculation (in this example, **squaring the number**).
  - Displays the result in the result text box.
  - If not successful,
    displays an error message in the result text box:

```
% Callback function for the calculate button
function calculate_callback(~,~)
    % Get the input from the text box
    input_str = get(input_box,'String');
    num = str2double(input_str);

    % Check if the input is a valid number
    if ~isnan(num)
              % If valid, perform calculation
  (example: square the number)
        result = num ^ 2;
        % Display the result
       set(result_box,'String',num2str(result));
   else
       % If not valid, display an error message
             set(result_box,'String','Invalid
 input. Please enter a number.');
              end
   end
```

The complete function (including the block codes described above) is given as follows:

Note: MATLAB App Designer automatically generates the required *function* and *end* code skeletons. Please paste the remaining code into the appropriate sections within these generated functions.

```
function CheckNumberApp
    % Create the UI components
    f = figure('Name','Number Checker App', 'Posi-
tion', [100 100 300 150]);

    % Text box for input
    input_box = uicon-
trol('Style','edit','String','','Position',[50 100
200 25]);

    % Button to trigger calculation
    calculate_button = uicontrol('Style','pushbut-
ton','String','Calculate','Position',[90 50 100
25],...
                                'Callback',@calcu-
late_callback);

    % Text box to display result
    result_box = uicon-
trol('Style','edit','String','','Position',[50 20 200
25],'Enable','inactive');

    % Callback function for the calculate button
    function calculate_callback(~,~)
        % Get the input from the text box
        input_str = get(input_box,'String');
        num = str2double(input_str);

        % Check if the input is a valid number
        if ~isnan(num)
        % If valid, perform calculation (example:
square the number)
        result = num ^ 2;
        % Display the result
        set(result_box,'String',num2str(result));
    else
        % If not valid, display an error message
        set(result_box,'String','Invalid input.
Please enter a number.');
    end
  end
end
```

Please do the following to use the App:

- Save the code as an M-file (e.g., **CheckNumberApp.m**).
- Run the M-file in MATLAB.
- Enter a number in the input text box.

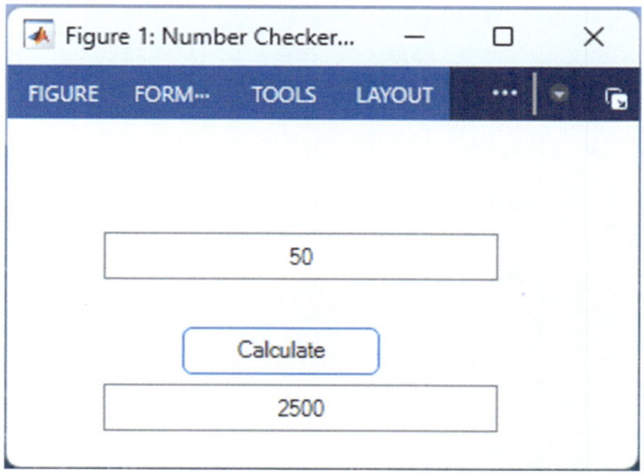

Fig. 6.1 Number Checker App

- Click the "Calculate" button (Fig. 6.1).

  The above App offers the following:

- **More precise Error Handling**: The code provides a specific error message to the user, making it more user-friendly.
- **Concise Code**: The code is well-structured and avoids unnecessary complexity.
- **Example Calculation**
- : The code includes a simple calculation (squaring the number) to demonstrate how to perform operations based on the input.

This app demonstrates basic input validation and calculations within a MATLAB App. You can easily adapt it to perform more complex calculations or incorporate additional input validation checks as needed.

6.3 Error Handling and Feedback

In MATLAB App Designer, Error Handling and Feedback are essential to making your App robust, user-friendly, and easy to debug. Error handling and feedback are very important for MATLAP Apps [7–9]. They provide

- Informative error messages to guide users.
- Implement graceful error handling to prevent unexpected behavior.

Here, we create an example that displays an error message when the user attempts to divide by zero. The process is described as follows:

1. **Create the Following UI Components**:

   - **figure**: Creates the main window of the App using UI—figure:

```
% Create the UI window
f = figure('Name', 'Divide by Zero App', 'Posi
tion', [100 100 350 220]);
```

- **uicontrol**: Creates the UI elements:
  - **edit**: create text boxes for the two input numbers (numerator and denominator) and create a text box to display the result.
  - **pushbutton**: create a button to trigger the division calculation:

```
% Labels and text boxes for input
uicontrol('Style', 'text', 'String', 'Enter Numera-
tor:', ...
              'Position', [30 160 120 25], 'Horizon-
talAlignment', 'left');
    num1_box = uicontrol('Style', 'edit', 'String',
'', ...
                         'Position', [160 160 150
25]);

    uicontrol('Style', 'text', 'String', 'Enter De-
nominator:', ...
              'Position', [30 120 120 25], 'Horizon-
talAlignment', 'left');
    num2_box = uicontrol('Style', 'edit', 'String',
'', ...
                         'Position', [160 120 150
25]);


% Button to trigger calculation
    calculate_button = uicontrol('Style', 'pushbut-
ton', 'String', 'Divide', ...
                                 'Position', [120 80
100 30], ...
                                 'Callback', @calcu-
late_callback);

% Result label + box
    uicontrol('Style', 'text', 'String', 'Result:',
...
              'Position', [30 40 120 25], 'Horizonta-
lAlignment', 'left');
    result_box = uicontrol('Style', 'edit', 'String',
'', ...
                           'Position', [160 40 150
25], 'Enable', 'inactive');
```

2. **Create a Calculate Callback Function**:

   - This function is executed when the "**Divide**" button is clicked. Here, we use a try-catch block to handle potential errors while computing division. The try-catch block lets you catch runtime errors and respond appropriately without crashing the App. So, the App
     - Gets the input values from the text boxes.
     - Performs the division operation.
     - Displays the result in the result text box.
     - catch block checks if the specific error is MATLAB—*divideByZero.* If so, it displays a specific error message. If it is a different error, it displays a generic error message.

The complete function (including the block codes described above) is given as follows:

```
function DivideByZeroApp
    % Create the UI window
    f = figure('Name', 'Divide by Zero App', 'Posi-
tion', [100 100 350 220]);

    % Labels and text boxes for input
    uicontrol('Style', 'text', 'String', 'Enter Nu-
merator:', ...
              'Position', [30 160 120 25], 'Horizon-
talAlignment', 'left');
    num1_box = uicontrol('Style', 'edit', 'String',
'', ...
                         'Position', [160 160 150
25]);

    uicontrol('Style', 'text', 'String', 'Enter De-
nominator:', ...
              'Position', [30 120 120 25], 'Horizon-
talAlignment', 'left');
    num2_box = uicontrol('Style', 'edit', 'String',
'', ...
                         'Position', [160 120 150
25]);

    % Button to trigger calculation
    calculate_button = uicontrol('Style', 'pushbut-
ton', 'String', 'Divide', ...
                                 'Position', [120 80
100 30], ...
                                 'Callback', @calcu-
late_callback);
```

```
    % Result label + box
    uicontrol('Style', 'text', 'String', 'Result:',
...
               'Position', [30 40 120 25], 'Horizonta-
lAlignment', 'left');
    result_box = uicontrol('Style', 'edit', 'String',
'', ...
                           'Position', [160 40 150
25], 'Enable', 'inactive');

    % Callback function for the calculate button
    function calculate_callback(~, ~)
        % Get input values
        num1 = str2double(get(num1_box, 'String'));
        num2 = str2double(get(num2_box, 'String'));

        % Check for valid numbers
        if isnan(num1) || isnan(num2)
            set(result_box, 'String', 'Invalid input.
Enter numbers only.');
            return;
        end

        % Handle division by zero
        if num2 == 0
            set(result_box, 'String', 'Error: Divi-
sion by zero!');
        else
            result = num1 / num2;
            set(result_box, 'String', num2str(re-
sult));
        end
    end
end
```

Upon entering the values (Fig. 6.2), the results are displayed in the results boxes (left) and an error due to division by zero (right).

This app demonstrates a simple example of handling division by zero within a MATLAB app. You can further enhance it by adding input validation (e.g., checking that input values are numbers) to improve the user experience.

User Preferences and Settings:

- Allow users to customize the app's behavior (e.g., color schemes, default values).
- Store user preferences for future sessions.

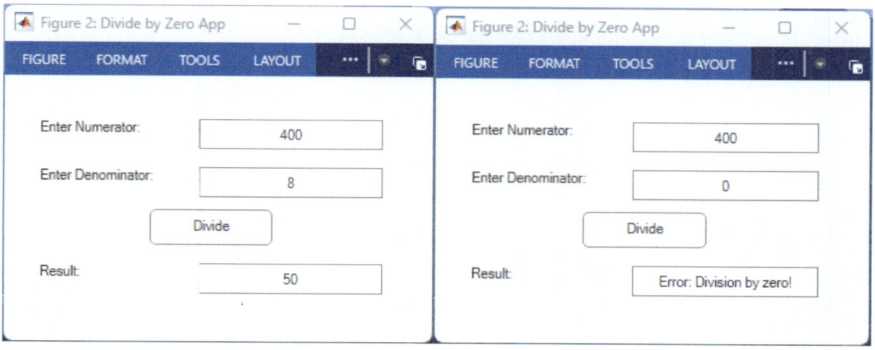

Fig. 6.2 Number Checker App: **a** A division example and **b** Division example with an error (divide by zero error)

6.3.1 Using **Uialert()** *for Critical Feedback*

In MATLAB, we can use `uialert()` to display a pop-up message in App Designer. Here we provide a simple example for *uialert* feedback. The whole process is explained as follows:

- **Open App Designer**:

In MATLAB **Command Window**, type *appdesigner*. Or you can click on Apps → Design Apps → + to create a new App.

- **Add a Panel**:

In the Component Library (left side), drag a Panel onto your app canvas and rename it to "**Simple UI Alert App**".

- **Add a UI Alert Push Button and call back**:

Drag a Push Button onto the panel and rename it to "**UI_Alert**" (Fig. 6.3). With the button selected, go to the Component Browser (bottom-right). Under Callbacks, click the "+" next to ButtonPushedFcn. App Designer will switch to Code View and create a function. Please add the following code (please note that App designer has already made a function skeleton and only add the necessary codes):

Note: MATLAB App Designer automatically generates the required *function* and *end* code skeletons. Please paste the remaining code into the appropriate sections within these generated functions.

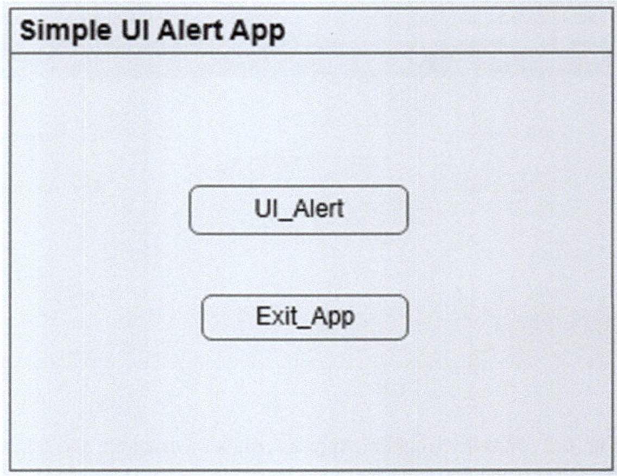

Fig. 6.3 Simple UI Alert App

```
function UI_AlertButtonPushed(app, event)
            % Display a simple alert message when the but-
ton is pressed
            uialert(app.UIFigure, ...
                'Hello! This is a simple uialert demonstra-
tion.', ...
                'Greetings', ...
                'Icon', 'info');
        end
```

- **Add an Exit App Push Button and call back**:

Drag a Push Button onto the panel and rename it to "**Exit_App**". With the button selected, go to the Component Browser (bottom-right). Under Callbacks, click the "+" next to ButtonPushedFcn. App Designer will switch to Code View and create a function. Please add the following code:

```
function Exit_AppButtonPushed(app, event)
     delete(app) % Exit App
 end
```

- **Save and Run App**:

Save the MATLAB App as "**Simple_UI_Alear.mlapp**". Click the **Run** ▶ button at the top of App Designer. The app window will open. Click the "UI_Aleart" button. A pop-up appears on your app window with the message (Fig. 6.4). Finally, click on "Exit_App" to close the App.

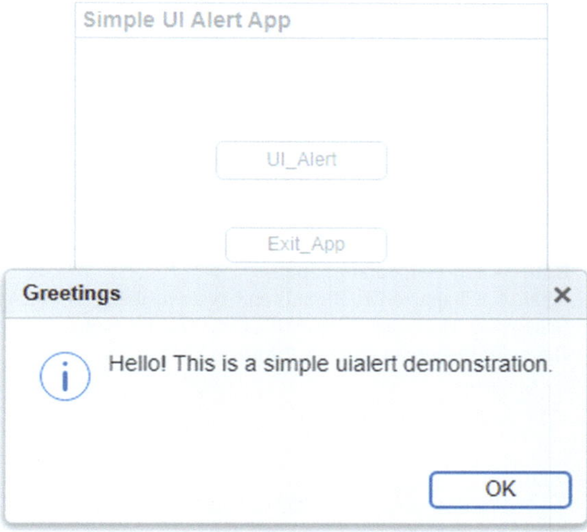

Fig. 6.4 Simple UI Alert App with message

6.4 Conclusion

In conclusion, this chapter has provided a comprehensive analysis of the crucial elements of handling User Input in your MATLAB applications, laying a foundation for creating robust, user-friendly Apps. In this chapter, we have covered why Data Validation and Input Refinement are both fundamental elements for maintaining data integrity and eliminating unexpected errors.

In addition, we have provided a thorough examination of why proper treatment of Error Handling and Feedback is required, including a visual pathway through error messages that is vital for improved User Experience. At last, we reviewed approaches toward User Preferences and settings.

Hence, mastering user input techniques, from rigorous input validation and proactive error management, is vital for developing sophisticated and reliable MATLAB Apps. By adopting the above-recommended model, you not only enhance the code's reliability but also improve the overall quality of the user interface experience, representing an initial step toward creating effective, engaging, and compelling interactive tools.

References

1. Valentine DT, Hahn BH. Essential MATLAB for engineers and scientists. Academic Press; 2022.
2. Hossain E. App Designer and Graphical User Interface in MATLAB. MATLAB and Simulink Crash Course for Engineers. Springer; 2022. p. 299–316.
3. Lent CS. Learning to program with MATLAB: Building GUI tools. John Wiley & Sons; 2022.

4. Mathworks: MATLAB App Designer: Create desktop and web apps in MATLAB. https://au.mathworks.com/products/matlab/app-designer.html (2025). Accessed September 2025.
5. Thabit AA, Alzidi E, Mohammed MA, Palchyk P. Development of a System Based on a Graphical User Interface Using MATLAB Tools. 2024 35th Conference of Open Innovations Association (FRUCT): IEEE; 2024. p. 744–54.
6. Darweesh M. Teaching MATLAB Programming by Utilizing Image Processing and Graphical User Interface Basics. Interactive Mobile Communication, Technologies and Learning. Springer; 2021. p. 824–32.
7. Attaway DC. Matlab: a practical introduction to programming and problem solving. 6th ed. Butterworth-Heinemann; 2022.
8. Akan A, Chaparro LF. Signals and systems using MATLAB®. Academic Press; 2024.
9. Schepers J, Hofmans J. TwoMP: A MATLAB graphical user interface for two-mode partitioning. Behavior research methods. 2009;41(2):507–14.

Part IV
Advanced App Development

Chapter 7
Advanced UI Components and Layouts

7.1 Introduction

Building sophisticated and interactive MATLAB applications goes beyond simple buttons and text fields. This chapter investigates advanced User Interface (UI) components and layout techniques available in MATLAB App Designer, which are essential for creating professional, scalable, and highly functional applications [1–3]. It covers the use of sophisticated UI components, such as tab groups, panels, tree controls, tables, and sliders, as well as dynamic content updates via callbacks. Emphasis is placed on effective layout management using grid layouts and flexible positioning strategies to ensure scalable, responsive App designs [4–6]. By the end of this chapter, you will be able to build complex, user-friendly MATLAB Apps that seamlessly integrate multiple interactive elements.

7.2 Advanced UI Components

While basic components, such as push buttons (*uibutton*), edit fields (*uieditfield*), and labels (*uilabel*), form the foundation, advanced components enable richer data display, complex interactions, and specialized visualizations [7, 8]. Advanced UI components, their features, and use cases are briefly outlined in Table 7.1.

Supplementary Information The online version contains supplementary material available at https://doi.org/10.1007/978-981-95-8797-1_7.

G. R. Naik, *MATLAB App Development*, https://doi.org/10.1007/978-981-95-8797-1_7

Table 7.1 Advanced UI components, features, and use cases

UI components	Purpose	Advanced features	Use case
Tables (`uitable`)	Perfect for presenting tabular data in a structured, scrollable format (matrices, cell arrays, or table objects)	Editable cells, sortable columns, editable column headers, row selection, and data update integration are some of the features	Displaying database queries, configuration settings, or experimental results
Trees (`uitree`, `uitreenode`)	Representing hierarchical data structures, similar to a file explorer or an object browser	Node selection callbacks, dynamic population, expandable or collapsible nodes, and unique node icons	Displaying project structures, navigating file systems, or creating organizational charts
Tab groups (`uitabgroup`, `uitab`)	Organizing several panels or sections of an application into a single, small area accessible via tabs is called a tab group (uitabgroup, uitab)	Each uitab serves as a container for additional UI elements, enabling content to be logically separated	Multi-page forms, various perspectives of the same data, or separating settings from primary features
Gauges and indicators (`uigauge`, `uiprogressdlg`)	Displaying data, status, or progress in real time	Dynamic progress dialogs and both linear and circular gauges	Displaying data quality, tracking simulation progress, or displaying sensor readings
Plots and axes (`uiaxes`)	Incorporating the robust plotting features of MATLAB straight into your application	Custom graphics, image display, 2D/3D plot support, and interactive plotting (panning, zooming)	Plotting signals in real time, visualizing data, or showing the outcomes of image processing
Custom components (using `uipanel` and nested components)	Assembling several fundamental components into a `UIPanel` or other container to create reusable, sophisticated UI elements	Logic and user interface encapsulation increase the modularity of app development	A specialized data viewer, a reusable control panel, or a custom data entry block

7.3 Advanced Layouts

Effective layout management is crucial for creating responsive, visually appealing MATLAB apps that adapt to different screen sizes while maintaining usability [7, 8]. Below, we outline commonly used UI Layouts, their features, and purpose. Their use cases are also briefly explained in Table 7.2.

Table 7.2 Advanced UI Layouts, features, and use cases

UI layouts	Purpose	Advanced features	Use case
Grid layouts (`uigridlayout`)	Organizing components in a grid that is adaptable (rows and columns). In current MATLAB Apps, this is the most effective and suggested layout manager	**Automatic Resizing**: When the App window size changes, components automatically adjust. **Row/Column Weights**: Indicate the distribution of space between rows and columns (e.g., Weight: "1x" for equal distribution, Weight: "fixed" for fixed size). **Padding and Spacing**: Manage the distance between grid boundaries and components**. Alignment**: Ensure that elements are precisely positioned inside their grid cells	Any complex design that calls for dynamic forms and dashboards, as well as exact alignment and responsiveness
Tab group layout (`uitabgroup`)	As mentioned above, uitabgroup not only organizes components into tabs but also acts as a layout container. Each uitab within the group is a separate container for other UI components, often using uigridlayout within each tab for further organization		Creating multi-page applications or organizing related functionalities into distinct sections

(continued)

Table 7.2 (continued)

UI layouts	Purpose	Advanced features	Use case
Panels (`uipanel`, `uitab`)	Grouping related UI components visually and logically. Panels are essential for breaking down complex interfaces into manageable sections	Can have titles, borders, and serve as containers for other layout managers (like uigridlayout)	Grouping input controls, displaying different data views, or creating distinct sections within a larger app
Responsive design consideration	Using uigridlayout with Weight: "1x" for rows/columns ensures that components scale proportionally when the app window is resized. Avoid using fixed pixel positions for components that require adaptability	While not a direct layout component, setting the MinimumWidth/ MinimumHeight properties on components or a uigridlayout can prevent elements from becoming too small to be usable	Dynamically showing or hiding components based on screen size or user preferences (e.g., using Visible: "on"/ "off")

Importance and Benefits

Utilizing advanced UI components and layout managers offers significant benefits:

- Enhanced User Experience (UX): Creates intuitive, organized, and visually appealing interfaces that are easy to navigate and understand.
- Scalability and Maintainability: Modular design with panels and custom components makes apps easier to extend, debug, and maintain.
- Responsiveness: Apps adapt gracefully to different screen resolutions and window sizes, improving usability across various devices.
- Data Visualization: Direct integration of MATLAB's plotting capabilities allows for robust and interactive data analysis within the app.
- Professional Appearance: Well-designed layouts and rich components give your MATLAB Apps a polished and professional look.

By strategically combining these advanced UI components with powerful layout managers, such as uigridlayout and uitabgroup, you can design and build highly sophisticated, interactive, and user-friendly MATLAB applications that meet complex requirements [7, 8].

Next, we explain how to use Advanced UI Components and Layouts in MATLAB Apps. Core components are described with examples.

7.4 Creating Custom UI Components

Here, we explain how to create custom UI components that help you extend the capabilities of built-in components by creating your own.

The following example (using sliders) provides a foundation for creating custom UI components in MATLAB. Here, we demonstrate the functionality of a slider as a UI component that displays its value upon click. The steps are outlined as follows (with the code):

1. **Create a Figure**:
   - Create the main window for the App using **uifigure**.
   - Set the title of the window using **Name**.
   - Set initial size and location of the window using **Position**:

```
% Create a figure
fig = uifigure('Name', 'Custom UI Components', 'Posi-
tion', [100 100 400 300]);
```

2. **Create a Custom Panel**:
   - Create a container within the figure using **uipanel**.
   - Set the title of the panel using **Title** and name it as "My Custom Panel".
   - Define the size and location of the panel within the figure using **Position**:

```
% Create a custom panel
panel = uipanel(fig, 'Title', 'My Custom Panel', 'Po-
sition', [50 50 300 200]);
```

3. **Create a Custom Button**:
   - Create a button within the panel using **uibutton**.
   - Set the text displayed on the button using the **Text** option.
   - Define the size and location of the button within the panel using Position.
   - Customize the button's appearance using **BackgroundColor** and **FontColor** options:

```
% Create a custom button
button = uibutton(panel, 'Text', 'Click Me!', 'Posi-
tion', [50 100 100 30]);
button.BackgroundColor = [0.8 0.8 0.8];
button.FontColor = 'blue';
```

4. **Add Button Callback**:

   - **ButtonPushedFcn** defines the function that will be executed when the button is clicked:

```
% Add a callback to the button
button.ButtonPushedFcn = @(~,~) disp('Button
clicked!');
```

5. **Create a Custom Slider**:

   - Create a slider within the panel using **uislider**.
   - Define the size and location of the slider using **Position**.
   - Set the minimum and maximum values for the slider using **Limits**.
   - Set the initial value of the slider using **Value**:

```
% Create a custom slider
slider = uislider(panel, 'Position', [50 50 200
20]);
slider.Limits = [0 100];
slider.Value = 50;
```

6. **Add Slider Callback**:

   - **ValueChangedFcn** defines the function that will be executed when the slider value changes. In this example, the current slider value is displayed in the command window:

```
% Add a callback to the slider
slider.ValueChangedFcn = @(~, event) disp(['Slider
value: ', num2str(event.Value)]);
```

The complete function (including the block codes described above) is given as follows.

Note: MATLAB App Designer automatically generates the required *function* and *end* code skeletons. Please paste the remaining code into the appropriate sections within these generated functions.

```
function CustomUIComponentsApp
% Create a figure
fig = uifigure('Name', 'Custom UI Components', 'Posi-
tion', [100 100 400 300]);

% Create a custom panel
panel = uipanel(fig, 'Title', 'My Custom Panel', 'Po-
sition', [50 50 300 200]);
```

```
% Create a custom button
button = uibutton(panel, 'Text', 'Click Me!', 'Posi-
tion', [50 100 100 30]);
button.BackgroundColor = [0.8 0.8 0.8];
button.FontColor = 'blue';
% Add a callback to the button
button.ButtonPushedFcn = @(~,~) disp('Button
clicked!');

% Create a custom slider
slider = uislider(panel, 'Position', [50 50 200 20]);
slider.Limits = [0 100];
slider.Value = 50;

% Add a callback to the slider
slider.ValueChangedFcn = @(~, event) disp(['Slider
value: ', num2str(event.Value)]);

end
```

To run the App:

1. Save the code as an M-file (e.g., CustomUIComponentsApp.m).
2. Run the M-file in MATLAB.

This will create a simple app with a custom panel, button, and slider. Upon clicking the button and moving the slider, we obtain the following result (Fig. 7.1).

After clicking the "Click Me!" button and moving the sliders, the following are displayed on the MATLAB workspace:

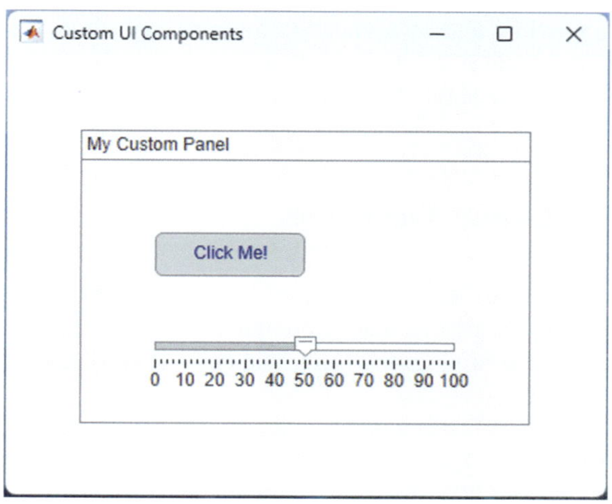

Fig. 7.1 UI component example using a slider

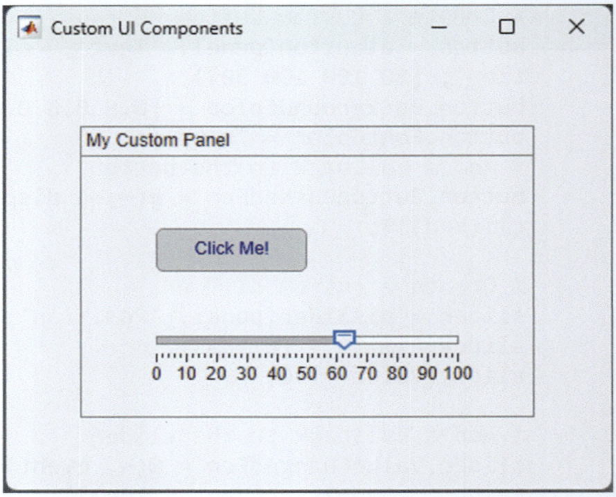

Fig. 7.2 UI Component example (when the slider is changed)

```
>> CustomUIComponentsApp
Button clicked!
Slider value: 62.5
```

We also outline the results of clicking the Button! Showing the following Slider value: 62.5 (Fig. 7.2).

You can modify this code further to

- Add more UI components (e.g., text boxes, dropdowns, images).
- Customize the appearance of the components (e.g., colors, fonts, sizes).
- Implement more complex interactions between components.
- Add data visualization elements (e.g., plots, charts).

This example provides a basic foundation for creating custom UI components in MATLAB. You can explore the documentation for uifigure, uipanel, and other UI components to learn more about their properties and methods.

7.5 Advanced Layout Techniques

UI components are used to build an advanced layout in MATLAB Apps. Advanced layout techniques can be used to master complex grid layouts, including spanning rows and columns and nesting grids. Here, you can use split panels and tab panels for complex layouts [7, 8].

Here, we create a MATLAB function named AdvancedLayoutApp that makes a simple MATLAB App using the uifigure, uigridlayout, and other UI components. Below, we outline the step-by-step procedure followed by the MATLAB implementation:

1. **Creating the Figure and Grid Layout**:

   - fig = uifigure(…): This line creates a new figure window with the specified title and dimensions. The figure window serves as the container for all the UI elements.
   - grid = uigridlayout(fig, [3, 2], …): This creates a grid layout within the figure, dividing the available space into 3 rows and 2 columns. The RowHeight and ColumnWidth properties specify the relative sizes of rows and columns:

   ```
   % Create a figure
   fig = uifigure('Name', 'Advanced Layout Techniques',
   'Position', [100 100 600 400]);
   % Create a grid layout
   grid = uigridlayout(fig, [3, 2], 'RowHeight', {'1x',
   '1x', '1x'}, 'ColumnWidth', {'1x', '1x'});
   ```

2. **Creating UI Components**:

   - edit1 = uieditfield(grid, 'text'): This creates an editable text field within the first cell of the grid.
   - edit1.Value = 'Edit Field 1': This sets the initial text displayed in the edit field.
   - Similar lines create another edit field (edit2), two buttons (button1 and button2), a listbox (listbox), and an axes object (axes1) for plotting:

   ```
   % Create components
   edit1 = uieditfield(grid, 'text');
   edit1.Value = 'Edit Field 1';
   edit2 = uieditfield(grid, 'text');
   edit2.Value = 'Edit Field 2';
   button1 = uibutton(grid, 'Text', 'Button 1');
   button2 = uibutton(grid, 'Text', 'Button 2');
   listbox = uilistbox(grid, 'Items', {'Item 1', 'Item
   2', 'Item 3'});
   axes1 = uiaxes(grid);
   ```

3. **Adding Data to the Axes**:

   - plot(axes1, rand(10,1)): This plots a random set of 10 data points on the axes:

   ```
   % Add data to axes
   plot(axes1, rand(10,1));
   ```

4. **Adjusting Component Sizes**:

   - grid.RowHeight{1} = '2x': This makes the first row of the grid twice as tall as the other rows.
   - grid.ColumnWidth{2} = '2x': This makes the second column of the grid twice as wide as the first column:

```
% Adjust component sizes
grid.RowHeight{1} = '2x'; % Make the first row taller
grid.ColumnWidth{2} = '2x'; % Make the second column
wider
```

The complete function (including the block codes described above) is given as follows:

```
function AdvancedLayoutApp
% Create a figure
fig = uifigure('Name', 'Advanced Layout Techniques',
'Position', [100 100 600 400]);
% Create a grid layout
grid = uigridlayout(fig, [3, 2], 'RowHeight', {'1x',
'1x', '1x'}, 'ColumnWidth', {'1x', '1x'});
% Create components
edit1 = uieditfield(grid, 'text');
edit1.Value = 'Edit Field 1';
edit2 = uieditfield(grid, 'text');
edit2.Value = 'Edit Field 2';
button1 = uibutton(grid, 'Text', 'Button 1');
button2 = uibutton(grid, 'Text', 'Button 2');
listbox = uilistbox(grid, 'Items', {'Item 1', 'Item
2', 'Item 3'});
axes1 = uiaxes(grid);
% Add data to axes
plot(axes1, rand(10,1));
% Adjust component sizes
grid.RowHeight{1} = '2x'; % Make the first row taller
grid.ColumnWidth{2} = '2x'; % Make the second column
wider
end
```

In essence, this App demonstrates a basic grid-based App in MATLAB. It showcases the use of various UI components and how to arrange them using the grid layout's properties.

To run the app:

1. Save the code as an M-file (**AdvancedLayoutApp.m**).
2. Run the M-file in MATLAB, and we get the following result (Fig. 7.3).

This example demonstrates some basic advanced layout techniques. You can further improve the layouts using the following: (i) Split Layout: divide the figure into two or more sections; (ii) Flow Layout: arrange components in a horizontal or vertical flow; and (iii) Using Padding and Spacing: adjust the spacing between components to create a visually appealing layout. By combining these techniques, you can create highly customized and user-friendly MATLAB Apps.

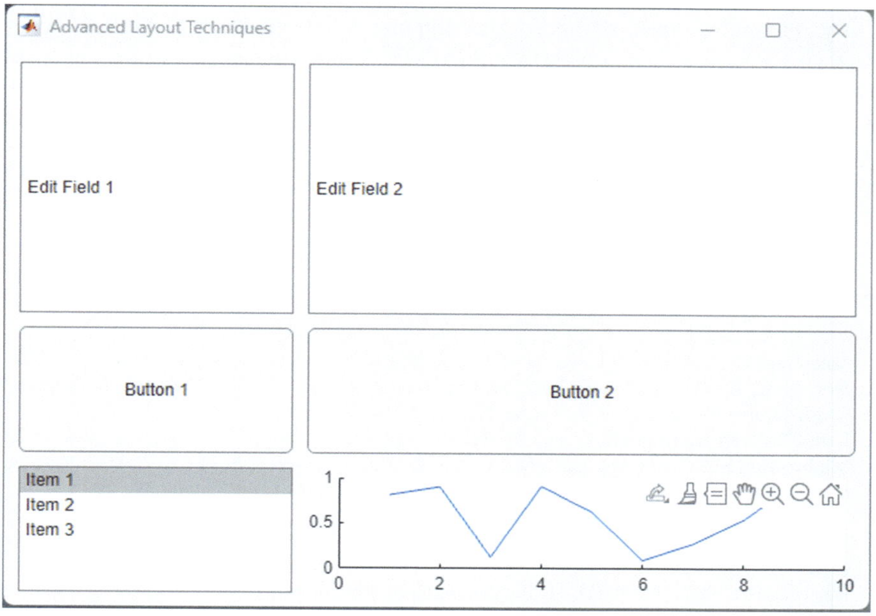

Fig. 7.3 Example of the advanced layout techniques

Advanced layout techniques are highly useful for successful App design in MATLAB. Some of the key advantages of advanced layout techniques include

- **Grid Layout**: Offers a flexible approach to arranging components in a grid-like structure.
- **Flexible Sizes**: Using "1x" or other flexible sizes allows components to resize proportionally as the window size changes.
- **Nested Layouts**: You can nest multiple layouts within each other to create more complex arrangements.
- **Using Position Property**: Precisely control the placement of components within the grid.

7.6 Conclusion

In conclusion, this chapter has provided a comprehensive overview of advanced UI components and layout techniques essential for developing sophisticated and interactive MATLAB applications. Importantly, we have investigated advanced layout management, emphasizing the pivotal role of uigridlayout for creating responsive and adaptable interfaces. The strategic use of uipanel for logical grouping and modular design was also highlighted, along with key considerations for responsive design, including fluid layouts and visibility control.

By integrating these advanced UI components and mastering effective layout strategies, you are now equipped to build MATLAB apps that not only offer enhanced user experiences through intuitive design and interactivity but also boast improved scalability, maintainability, and a truly professional appearance. These techniques are fundamental for transforming complex functionality into user-friendly, robust applications.

References

1. Valentine DT, Hahn BH. Essential MATLAB for engineers and scientists. Academic Press; 2022.
2. Hossain E. App Designer and Graphical User Interface in MATLAB. MATLAB and Simulink Crash Course for Engineers. Springer; 2022. p. 299–316.
3. Lent CS. Learning to program with MATLAB: Building GUI tools. John Wiley & Sons; 2022.
4. Mathworks: MATLAB App Designer: Create desktop and web apps in MATLAB. https://au.mathworks.com/products/matlab/app-designer.html (2025). Accessed September 2025.
5. Thabit AA, Alzidi E, Mohammed MA, Palchyk P. Development of a System Based on a Graphical User Interface Using MATLAB Tools. 2024 35th Conference of Open Innovations Association (FRUCT): IEEE; 2024. p. 744–54.
6. Darweesh M. Teaching MATLAB Programming by Utilizing Image Processing and Graphical User Interface Basics. Interactive Mobile Communication, Technologies and Learning. Springer; 2021. p. 824–32.
7. Attaway DC. Matlab: a practical introduction to programming and problem solving. 6th ed. Butterworth-Heinemann; 2022.
8. Akan A, Chaparro LF. Signals and systems using MATLAB®. Academic Press; 2024.

Chapter 8
Interactive Visualizations

8.1 Introduction

Interactive visualizations play an essential role in modern data analysis and App development because they transform static data into dynamic, exploratory experiences [1–3]. This chapter explores the interactive visualization applications within the MATLAB environment. Here, we explore how to leverage MATLAB's powerful visualization capabilities to create dynamic and engaging graphical representations of user data.

In MATLAB, interactive visualizations enable researchers, engineers, and students to test hypotheses instantly, observe outcomes visually, and refine models without rerunning code repeatedly [4–6]. They also make Apps more engaging and user-friendly, bridging the gap between complex computations and intuitive graphical interpretation. Interactive visualizations go beyond static plots, offering users the ability to

- **Explore data interactively**: Zoom, pan, and rotate plots to gain deeper insights.
- **Control parameters**: Adjust variables and observe their effects on the visualization in real time.

By incorporating interactivity, you can transform your data analysis from a passive experience into an active and insightful exploration. In this chapter, we cover the three crucial data visualization concepts, which include *i) Dynamic plots, (ii) Zoom and Pan, and (iii) Data cursors. By the end of this chapter, you will have acquired the knowledge and skills to create compelling, insightful interactive visualizations that effectively communicate your data and facilitate a more profound understanding.

Supplementary Information The online version contains supplementary material available at https://doi.org/10.1007/978-981-95-8797-1_8.

G. R. Naik, *MATLAB App Development*, https://doi.org/10.1007/978-981-95-8797-1_8

8.2 Dynamic Plots

Dynamic plots are MATLAB Apps that display values and update them in real time in response to user input [7, 8]. Here, we create a plot that updates slider value changes. Below we provide a step-by-step explanation:

1. **Create Figure and Grid Layout**:
   - Create a main window for the app using **uifigure**.
   - Divide the figure into two rows: one for the slider and one for the plot using **uigridlayout**:

```
% Create a figure
fig = uifigure('Name', 'Dynamic Plot', 'Position',
[100 100 400 300]);

% Create a grid layout
grid = uigridlayout(fig, [2, 1]);
grid.RowHeight = {'1x', '3x'};
```

2. **Create Slider**:
   - Create a slider component using **uislider**.
   - Set the minimum and maximum values of the slider using **Limits**:

```
% Create a slider
slider = uislider(grid);
slider.Limits = [0 10];
slider.Value = 5;
```

3. **Create Axes**:
   - Create an axes object for plotting using **uiaxes**:

```
% Create axes
ax = uiaxes(grid);
```

4. **Initialize Plot Data**:
   - Initialize x and y values for the plot:

```
% Initialize plot data
x = linspace(0, 10, 100);
y = sin(x);
```

5. **Create Initial Plot**:
   - Create the initial plot on the axes using **plot**:

```
% Create the initial plot
plot(ax, x, y);
ax.Title.String = 'Dynamic Sine Wave';
```

6. Update Plot Callback:

   - Use **ValueChangedFcn** callback, a function that is triggered when the slider value changes.
   - Use the **updatePlot** function to update the plot based on the new slider value:

```
% Update Plot Callback
  slider.ValueChangedFcn = @(src, event) up-
  datePlot(event.Value, ax);
```

7. **updatePlot Function**:

   - It updates the y data based on the sliderValue.
   - **cla**(axesHandle) clears the previous plot from the axes.
   - **plot** creates a new plot with updated data:

```
function updatePlot(sliderValue, ax)
    % Update data based on slider value
    x = linspace(0, 10, 100);
    y = sin(x + sliderValue);

    % Clear and update plot
    cla(ax);
    plot(ax, x, y);
end
```

Key Features:

- **Dynamic Updates**: The plot updates in real time as the slider value is changed.
- **Slider Control**: The slider provides an interactive way to control the plot.

The complete function (including the block codes described above) is given as follows:

Note: MATLAB App Designer automatically generates the required *function* and *end* code skeletons. Please paste the remaining code into the appropriate sections within these generated functions.

```
function DynamicPlotApp
    % Create a figure
    fig = uifigure('Name', 'Dynamic Plot', 'Position', [100
100 400 300]);

    % Create a grid layout
```

```
    grid = uigridlayout(fig, [2, 1]);
    grid.RowHeight = {'1x', '3x'};

    % Create a slider
    slider = uislider(grid);
    slider.Limits = [0 10];
    slider.Value = 5;

    % Create axes
    ax = uiaxes(grid);

    % Initialize plot data
    x = linspace(0, 10, 100);
    y = sin(x);

    % Create the initial plot
    plot(ax, x, y);
    ax.Title.String = 'Dynamic Sine Wave';

    % Assign callback
    slider.ValueChangedFcn = @(src, event) up-
datePlot(event.Value, ax);
end

function updatePlot(sliderValue, ax)
    % Update data based on slider value
    x = linspace(0, 10, 100);
    y = sin(x + sliderValue);

    % Clear and update plot
    cla(ax);
    plot(ax, x, y);
end
```

To Run This App:

1. Save the code as an M-file (e.g., `DynamicPlotApp.m`).
2. Run the M-file in MATLAB.

This will create a simple app with a dynamic plot that responds to slider movements (Fig. 8.1).

You can further enhance this app by (i) adding more controls (e.g., buttons, drop-downs) to modify plot parameters; (ii) implementing different plot types (e.g., bar plot, scatter plot); (iii) adding animations to the plot; and iv) using more complex data for plotting. This example provides a solid foundation for creating dynamic and interactive plots in MATLAB.

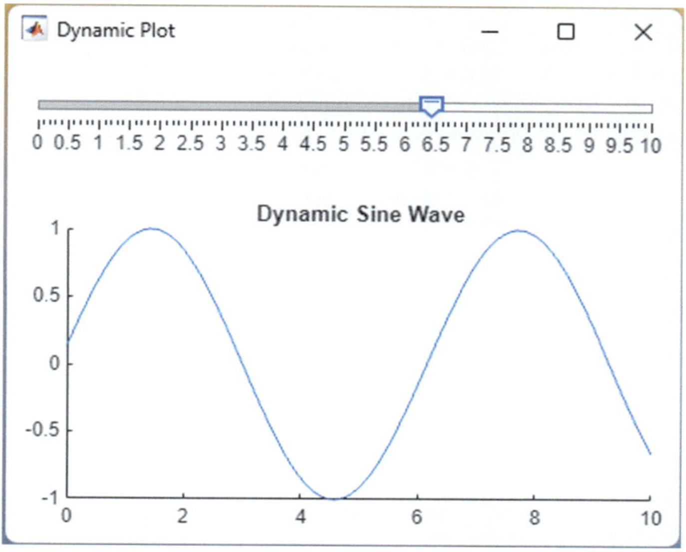

Fig. 8.1 Dynamic plot App (slider bar—signal value changes) example

8.3 Zoom and Pan

Built-in tools, such as the Zoom and Pan tools in MATLAB, allow users to zoom in, zoom out, and interactively move plots [7, 8]. In addition, users can modify or build Zoom and Pan Apps to create customized applications. The following example enables users to Zoom and Pan on plots for detailed exploration. This example also demonstrates the basic principles of implementing Zoom and Pan functionality in a MATLAB app. The App enables users to Zoom and Pan on plots for detailed exploration. This helps users to adapt and extend these techniques to create more sophisticated and interactive visualizations. The App creation is explained as follows:

1. **Create Figure**:

   - Create a main figure window for the app using `figure()`:

   ```
   % Create the figure
   f = figure('Name', 'Zoom and Pan App', 'Position', [100
   100 600 400]);
   ```

2. **Generate Sample Data**:

   - Sample data, `x` and `y`, are created for a simple sine wave example. You can replace this with your own data:

   ```
   % Generate sample data
   x = linspace(0, 10, 100);
   y = sin(x) + rand(1, 100);
   ```

3. **Create Plot**:
   - Create a set of axes within the figure using `axes()`.
   - Create the line plot of the data using `plot()`.
   - Add grid lines to the plot for better readability using `grid()`:

```
% Create the plot
ax = axes('Parent', f);
plot(ax, x, y);
grid(ax, 'on');
```

4. **Enable Zoom and Pan**:
   - Enable the zoom functionality for the axes using `zoom(ax, 'on')` option.
   - Enable the pan functionality for the axes using `pan(ax, 'on')` option:

```
% Create zoom and pan objects
z = zoom(f);
p = pan(f);
```

5. **Add UI Controls (Optional)**:
   - Create pushbuttons for "Zoom In," "Zoom Out," "Pan On," "Pan Off," and "Reset Zoom." Each button has a corresponding callback function that interacts with the axes:

```
% Add UI controls
uicontrol('Parent', f, 'Style', 'pushbutton', ...
              'String', 'Zoom In', 'Position', [20
350 80 25], ...
              'Callback', @(~,~)
zoom_in_callback(z));
    uicontrol('Parent', f, 'Style', 'pushbutton', ...

              'String', 'Zoom Out', 'Position', [110
350 80 25], ...
              'Callback', @(~,~)
zoom_out_callback(z));
    uicontrol('Parent', f, 'Style', 'pushbutton', ...
              'String', 'Pan On', 'Position', [20 300
80 25], ...
              'Callback', @(~,~) pan_on_callback(p));
    uicontrol('Parent', f, 'Style', 'pushbutton', ...
              'String', 'Pan Off', 'Position', [110
300 80 25], ...
              'Callback', @(~,~)
```

```
pan_off_callback(p));
    uicontrol('Parent', f, 'Style', 'pushbutton', ...
              'String', 'Reset Zoom', 'Position', [20
250 80 25], ...
              'Callback', @(~,~) re-
set_zoom_callback(ax));
```

6. **Callback Functions**: Use the following callback functions:

- `zoom_in_callback()`: Zooms in on the current plot.
- `zoom_out_callback()`: Zooms out on the current plot.
- `pan_on_callback()`: Enables panning mode.
- `pan_off_callback()`: Disables panning mode.
- `reset_zoom_callback()`: Resets the axis limits to their initial values:

```
% Callback functions
   function zoom_in_callback(~)
       set(z, 'Enable', 'on');
       zoom(1.2); % Zoom in by 20%
   end

   function zoom_out_callback(~)
       set(z, 'Enable', 'on');
       zoom(0.8); % Zoom out by 20%
   end

  function pan_on_callback(~)
        set(p, 'Enable', 'on');
  end

  function pan_off_callback(~)
        set(p, 'Enable', 'off');
  end

  function reset_zoom_callback(ax)
       axis(ax, 'auto'); % Reset axes
  end
```

The complete function (including the block codes described above) is given as follows.

Note: MATLAB App Designer automatically generates the required *function* and *end* code skeletons. Please paste the remaining code into the appropriate sections within these generated functions.

```
function zoom_pan_app
    % Create the figure
    f = figure('Name', 'Zoom and Pan App', 'Posi-
tion', [100 100 600 400]);

    % Generate sample data
    x = linspace(0, 10, 100);
    y = sin(x) + rand(1, 100);

    % Create the plot
    ax = axes('Parent', f);
    plot(ax, x, y);
    grid(ax, 'on');

    % Create zoom and pan objects
    z = zoom(f);
    p = pan(f);

    % Add UI controls
    uicontrol('Parent', f, 'Style', 'pushbutton', ...
              'String', 'Zoom In', 'Position', [20
350 80 25], ...
              'Callback', @(~,~)
zoom_in_callback(z));
    uicontrol('Parent', f, 'Style', 'pushbutton', ...
              'String', 'Zoom Out', 'Position', [110
350 80 25], ...
              'Callback', @(~,~)
zoom_out_callback(z));
    uicontrol('Parent', f, 'Style', 'pushbutton', ...
              'String', 'Pan On', 'Position', [20 300
80 25], ...
              'Callback', @(~,~) pan_on_callback(p));
```

```
    uicontrol('Parent', f, 'Style', 'pushbutton', ...
              'String', 'Pan Off', 'Position', [110
300 80 25], ...
              'Callback', @(~,~)
pan_off_callback(p));
    uicontrol('Parent', f, 'Style', 'pushbutton', ...
              'String', 'Reset Zoom', 'Position', [20
250 80 25], ...
              'Callback', @(~,~) re-
set_zoom_callback(ax));

    % Callback functions
    function zoom_in_callback(~)
        set(z, 'Enable', 'on');
        zoom(1.2); % Zoom in by 20%
    end

    function zoom_out_callback(~)
        set(z, 'Enable', 'on');
        zoom(0.8); % Zoom out by 20%
    end

    function pan_on_callback(~)
        set(p, 'Enable', 'on');
    end

    function pan_off_callback(~)
        set(p, 'Enable', 'off');
    end

    function reset_zoom_callback(ax)
        axis(ax, 'auto'); % Reset axes
    end
end
```

To Use This App:

1. Save the code as an `.m` file (e.g., `zoom_pan_app.m`).
2. Run the script in MATLAB. This will create a figure with the plot and the UI controls (Fig. 8.2). You can use this to (i) use the mouse to zoom in/out and pan around the plot, and (ii) click the buttons to control zoom and pan programmatically.

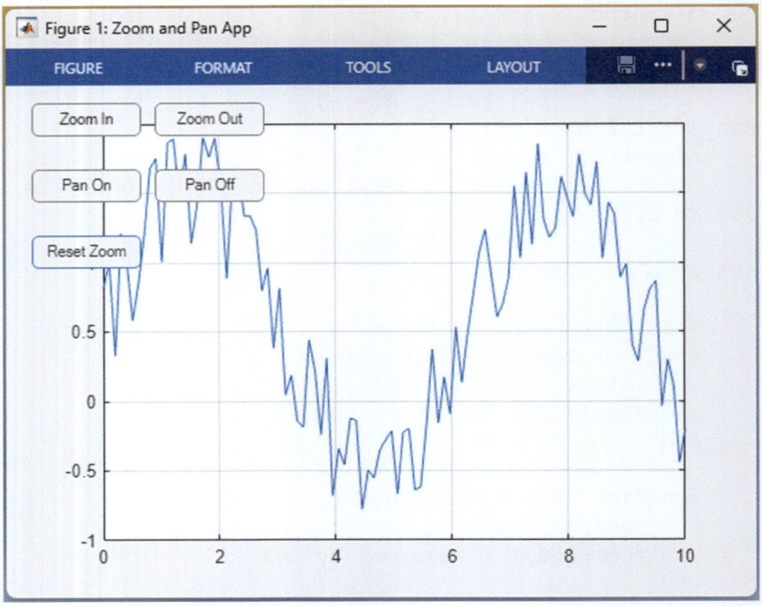

Fig. 8.2 Zoom Pan App (Zoom in, Zoom out, Pan On, Pan off, and Reset Zoom options) example

This App provides a basic framework for a zoom-and-pan app. You can customize it further by

- Adding more UI controls (e.g., for saving the plot, changing plot properties).
- Implementing more advanced zoom/pan features (e.g., zooming to a specific region).
- Integrating with other MATLAB features (e.g., data cursors, linked plots).

8.4 Data Cursors

Data Cursors display detailed information about specific data points on a plot. Data cursors are essential interactive visualization tools in MATLAB used for a wide range of applications [7, 8]. Here, we describe a MATLAB App that displays detailed information about specific data points on a plot. The whole process is explained as follows:

1. **Create Figure, Sample Data, and Plot:**
   - Creates a figure, sample data, and plots for the sample data:

```
% Create the figure
f = figure('Name', 'Data Cursor App', 'Position',
[100 100 600 400]);
% Generate sample data
x = linspace(0, 10, 100);
y = sin(x) + rand(1, 100);

% Create the plot
ax = axes('Parent', f);
plot(ax, x, y);
grid(ax, 'on');
title('Data Cursor App');
```

2. **Create Data Cursor Mode Object:**
   - Create a data cursor mode object associated with the figure using `datacursormode(f)`:

```
% Create data cursor mode object
dcm_obj = datacursormode(f);
```

3. **Customize Data Tip Appearance (Optional):**
   - Set the display style for the data tip (e.g., "window" for a separate window, "datatip" for a tooltip) using `DisplayStyle`.
   - Determine if the cursor snaps to the nearest data point using `SnapToDataVertex`:

```
% Customize data tip appearance
set(dcm_obj, 'DisplayStyle', 'window'); % Display
data in a separate window
% set(dcm_obj, 'SnapToDataVertex', 'on'); % Snap to
nearest data point
```

4. **Update Data Tip Text Function:**
   - Define a custom function to update the text displayed in the data tip using `set(dcm_obj,'UpdateFcn', {@my_datacursor_updatefcn, x, y})`.
   - `my_datacursor_updatefcn` takes the data cursor object, event object, and the original data (x, y) as input. It finds the closest data point to the cursor position and creates a custom text string with the X and Y values:

```
% Update data tip text function
set(dcm_obj, 'UpdateFcn', {@my_datacursor_updatefcn,
x, y});
function output_txt = my_datacursor_updatefcn(obj,
event_obj, x, y)
    % Get position of data cursor
    pos = get(event_obj, 'Position');

    % Find the closest data point index
    [~, idx] = min(abs(x - pos(1)));

    % Create custom data tip text
    output_txt = {['X: ', num2str(x(idx), '%.2f')],
...
                  ['Y: ', num2str(y(idx), '%.2f')]};
end
```

The complete function (including the block codes described above) is given as follows:

```
function data_cursor_app
% Create the figure
f = figure('Name', 'Data Cursor App', 'Position',
[100 100 600 400]);
% Generate sample data
x = linspace(0, 10, 100);
y = sin(x) + rand(1, 100);

% Create the plot
ax = axes('Parent', f);
plot(ax, x, y);
grid(ax, 'on');
title('Data Cursor App');

% Create data cursor mode object
dcm_obj = datacursormode(f);

% Customize data tip appearance
set(dcm_obj, 'DisplayStyle', 'window'); % Display
data in a separate window
% set(dcm_obj, 'SnapToDataVertex', 'on'); % Snap to
nearest data point
```

```
% Update data tip text function
set(dcm_obj, 'UpdateFcn', {@my_datacursor_updatefcn,
x, y});

end

function output_txt = my_datacursor_updatefcn(obj,
event_obj, x, y)
    % Get position of data cursor
    pos = get(event_obj, 'Position');

    % Find the closest data point index
    [~, idx] = min(abs(x - pos(1)));

    % Create custom data tip text
    output_txt = {['X: ', num2str(x(idx), '%.2f')],
...
                  ['Y: ', num2str(y(idx), '%.2f')]};
end
```

To Use This App:

1. Save the code as an `.m` file (e.g., `data_cursor_app.m`).
2. Run the script in MATLAB.
3. Click on the plot to create data cursors. The data tips will display the X and Y coordinates of the selected points (Fig. 8.3).

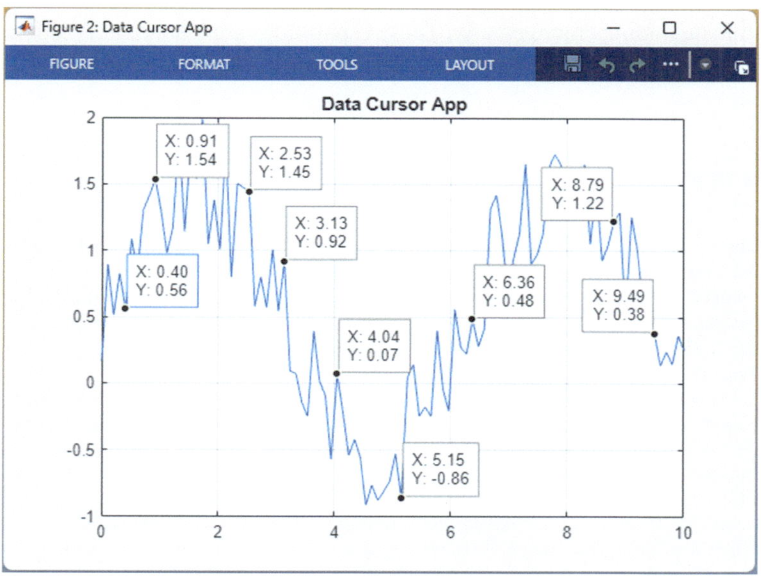

Fig. 8.3 Data cursor example App

This example demonstrates the data cursor example App. You can customize data tips by modifying the `my_datacursor_updatefcn` function to display any desired information, such as additional data variables, calculated values, and formatted text. Additionally, customize the appearance of the data tips using the properties of the `DataCursorMode` object. This example provides a basic framework for creating a data cursor app in MATLAB. You can further enhance it by (i) adding more data points and variables, (ii) implementing interactive features (e.g., deleting data cursors, changing data tip colors), an d iii) integrating with other MATLAB features (e.g., zooming, panning).

8.5 Conclusion

In this chapter, we explored the necessary Apps and tools for creating interactive visualizations that enhance data analysis and interpretation. First, we examined dynamic plots to update visualizations in real time, enabling users to observe data and model changes more effectively. We then discussed Zooming and Panning, which provide spontaneous navigation within complex datasets, making it easier to focus on regions of interest without losing context. Finally, we introduced data cursors, a powerful feature that enables precise inspection of individual data points and offers valuable insights beyond what static plots can provide.

Together, these interactive methods transform static graphics into engaging, user-driven evaluation tools. By integrating dynamic updates, navigational control, and point-specific information, analysts can discover hidden patterns, verify results, and communicate findings more clearly. These capabilities are essential in modern data-driven environments, where flexibility, precision, and clarity are essential.

References

1. Hossain E. App Designer and Graphical User Interface in MATLAB. MATLAB and Simulink Crash Course for Engineers. Springer; 2022. p. 299–316.
2. Mathworks: MATLAB App Designer: Create desktop and web apps in MATLAB. https://au.mathworks.com/products/matlab/app-designer.html (2025). Accessed September 2025.
3. Schepers J, Hofmans J. TwoMP: A MATLAB graphical user interface for two-mode partitioning. Behavior research methods. 2009;41(2):507-14.
4. Valentine DT, Hahn BH. Essential MATLAB for engineers and scientists. Academic Press; 2022.
5. Lent CS. Learning to program with MATLAB: Building GUI tools. John Wiley & Sons; 2022.
6. Thabit AA, Alzidi E, Mohammed MA, Palchyk P. Development of a System Based on a Graphical User Interface Using MATLAB Tools. 2024 35th Conference of Open Innovations Association (FRUCT): IEEE; 2024. p. 744–54.
7. Darweesh M. Teaching MATLAB Programming by Utilizing Image Processing and Graphical User Interface Basics. Interactive Mobile Communication, Technologies and Learning. Springer; 2021. p. 824–32.
8. Attaway DC. Matlab: a practical introduction to programming and problem solving. 6th ed. Butterworth-Heinemann; 2022.

Chapter 9
Performance Optimization

9.1 Introduction

MATLAB is commonly used for numerical computing, data analysis, and algorithm development. Although its high-level programming environment makes development easier, an inefficient implementation significantly hampers performance, especially with large datasets or complex simulations. In general, performance optimization in MATLAB aims to enhance execution speed, reduce memory usage, and maximize hardware utilization [1, 2].

Apps built using App Designer programmatically (such as *uifigure* and *uicontrol*) most often involve coupling an interactive User Interface (UI) with algorithms, including signal processing, computationally intensive tasks, data visualization, and, in some cases, simulations. Poorly optimized code written in MATLAB may result in Apps (i) being non-responsive, resulting in laggy UI, and slow button clicks and (ii) consuming a lot of unwanted memory, which may cause the App to become non-responsive, and (iii) the developer ends up designing and developing such Apps that are unprofessional for end-users. Therefore, effective performance optimization principles will ensure that MATLAB Apps stay **responsive, scalable,** and **efficient**, even when handling large datasets or when their operations are complex [3–6].

This chapter discusses the performance optimization problems in the MATLAB App environment. We will explore how optimization can enhance MATLAB App development through *profiling*, *caching*, and related techniques, and illustrate key strategies with examples.

Supplementary Information The online version contains supplementary material available at https://doi.org/10.1007/978-981-95-8797-1_9.

G. R. Naik, *MATLAB App Development*, https://doi.org/10.1007/978-981-95-8797-1_9

9.2 Key Optimization Strategies in App Development

Optimization strategies are needed for better App development in MATLAB. This section covers various optimization methods, supported by MATLAB examples and Apps.

9.2.1 Responsive UI with Efficient Code

It is essential to applying the most effective UI strategies and coding practices when designing MATLAB Apps. For example, in the App's Callback Functions (e.g., *ButtonPushedFcn*, *ValueChangeFcn*), avoid slow loops or unnecessary calculations. Instead of updating the points individually in a loop, it is vital to vectorize the plot in the plot App [4, 6]. For example, the code below illustrates both good (quicker and more efficient) and bad (slow) practices:

```
% Bad (slows the overall process):
for i = 1:length(x)
    plot(app.UIAxes, x(1:i), y(1:i));
    drawnow;
end

% Good (quicker and more efficient):
plot(app.UIAxes, x, y);
```

9.2.2 Preallocate Data

It is essential to assessing data and processing requirements during the design of any MATLAB application tool. To prevent delays (lags) when you submit a series of operation (repeated operation) requests, you may consider *preallocating arrays* in your App that process incoming data (for example, sensor data and/or files) [4, 6]. The code below illustrates a *preallocation* example [4, 6]:

```
data = rand(1e8,1); % large dataset
result = zeros(size(data)); % preallocate
```

9.2.3 *Efficient Plotting and Visualization*

For interactive MATLAB Apps with a large amount of data, the best way to implement the solutions is as follows:

- Use Plot or Scatter and then update the XData and YData properties rather than replotting.
- Use *downsampling* to allow UI responsiveness if the data is too large.

Here is a straightforward example that updates the current line rather than replotting, as shown in the following code:

```
% Update existing line instead of re-plotting
app.PlotLine.XData = newX;
app.PlotLine.YData = newY;
```

9.2.4 *Logical Indexing and Parallel Computing*

For improved MATLAB App development, logical indexing optimization techniques are required. Logical indexing makes it easy to filter data quickly in an application because it is faster than looping. However, parallel computing is helpful because tasks can be run simultaneously. An example of this involves using background workers **parfor** or **parfeval** to ensure the GUI remains responsive even as the tasks are being undertaken in parallel [4, 6]. An example of parallel computing involves performing image filtering (in the background) within an application without interrupting operations in the menus (interacting with menus).

9.2.5 *Memory Management in Long-Running Apps*

Sometimes, certain MATLAB Apps are left open for hours or even days. Then, instead of unnecessary copying, you can **delete** (clear) **vars** or **overwrite** your variables. In addition, sparse matrices are employed to reduce memory usage for large sparse data, such as graphs or adjacency matrices [4, 6].

9.2.6 Example: Performance Optimization—Real-Time App

Here, we present an example MATLAB App Designer that demonstrates how to optimize performance in a real-time plotting application. This application effectively updates plots, simulates a data stream, and maintains a responsive interface through *preallocation* and effective UI updates. Below, we describe the code, plot, and optimization techniques used:

```
classdef RealTimePlotApp < matlab.apps.AppBase
    % App components
    properties (Access = public)
        UIFigure        matlab.ui.Figure
        UIAxes          matlab.ui.control.UIAxes
        StartButton     matlab.ui.control.Button
        StopButton      matlab.ui.control.Button
    end

    % Internal properties
    properties (Access = private)
        TimerObj       timer
        DataBuffer     double
        BufferSize     = 1000  % Preallocated buffer size
        Index          = 1
        PlotLine       matlab.graphics.chart.primitive.Line
    end

    methods (Access = private)

        % Start data streaming
        function startStreaming(app)
            % Preallocate buffer for performance
            app.DataBuffer = zeros(1, app.BufferSize);
            app.Index = 1;

            % Timer for real-time updates
            app.TimerObj = timer( ...
                'ExecutionMode','fixedRate', ...
                'Period',0.05, ...    % update every 50 ms
                'TimerFcn',@(~,~)app.updatePlot);

            start(app.TimerObj);
        end

        % Stop data streaming
        function stopStreaming(app)
            if ~isempty(app.TimerObj) && isvalid(app.TimerObj)
                stop(app.TimerObj);
```

```
                delete(app.TimerObj);
            end
        end

        % Update plot efficiently
        function updatePlot(app)
            % Simulate new data point (e.g., sensor reading)
            newData = sin(0.1*app.Index) + 0.1*randn;

            % Store in buffer
            app.DataBuffer(app.Index) = newData;
            app.Index = app.Index + 1;

            % Reset index when buffer is full (rolling plot)
            if app.Index > app.BufferSize
                app.Index = 1;
            end
                 % Update existing line (faster than replotting)
            set(app.PlotLine, 'YData', app.DataBuffer);
        end
    end

    % App initialization and UI components
    methods (Access = private)

        function createComponents(app)
            % Create UIFigure
            app.UIFigure = uifigure('Position',[100 100 600 400]);

            % Create UIAxes
            app.UIAxes = uiaxes(app.UIFigure, ...
                'Position',[50 100 500 250]);
            xlabel(app.UIAxes,'Time Index');
            ylabel(app.UIAxes,'Signal');
            title(app.UIAxes,'Real-Time Data Stream');

            % Create buttons
            app.StartButton = uibutton(app.UIFigure,'push', ...
                'Text','Start', ...
                'Position',[150 30 100 40], ...
                'ButtonPushedFcn', @(~,~)app.startStreaming);

            app.StopButton = uibutton(app.UIFigure,'push', ...
                'Text','Stop', ...
                'Position',[350 30 100 40], ...
                'ButtonPushedFcn', @(~,~)app.stopStreaming);

          % Precreate plot line for efficient updating
      app.PlotLine = plot(app.UIAxes, nan(1,app.BufferSize), 'b-');
        end
    end
```

```
        methods (Access = public)
            % Constructor
            function app = RealTimePlotApp
                createComponents(app);
                  registerApp(app, app.UIFigure);
            end
        end
    end
```

In this example, the plot simulates the data until the Stop button is clicked (Fig. 9.1). In this scenario, we use the following optimization strategies:

- **Preallocation**: To avoid dynamic growth, DataBuffer is preallocated with zeros (BufferSize = 1000).
- **Effective plotting**: We only update YData (set (app.PlotLine, 'YData', ...)) after creating PlotLine once rather than re-plotting.
- **Timer object**: Handles user input and maintains responsiveness by dealing with real-time updates without affecting the user interface.
- **Rolling buffer**: It prevents the memory from growing endlessly whenever it is continuously streaming.

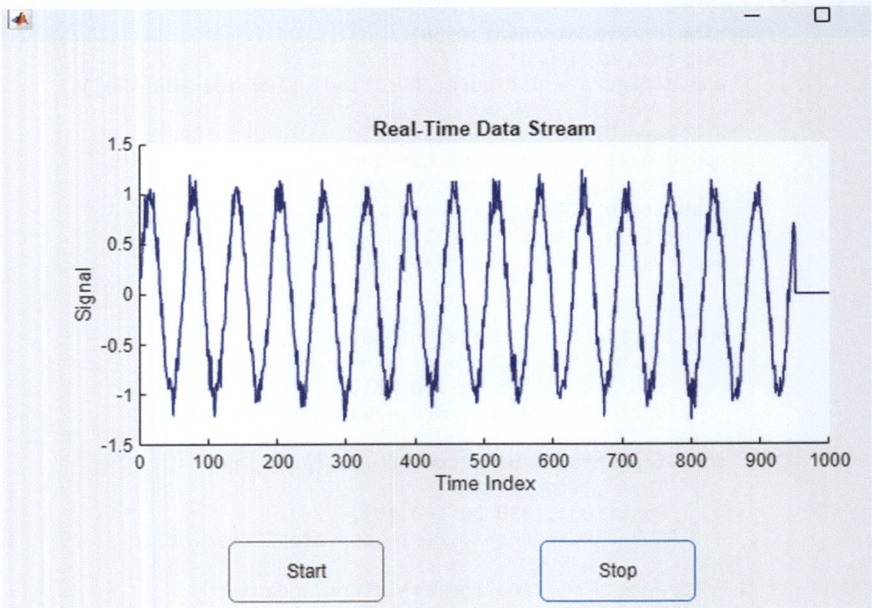

Fig. 9.1 Performance optimization example—real-time: Start and Stop options

9.2.7 *Example: Performance Optimization—Real-Time App with Background Computation*

We extend the real-time plotting application by adding parallel background computation, such as filtering a signal or extracting its features, as in the earlier example. Still, real-time data plotting continues in the modified (extended) application (on the main UI thread). In this example, it uses **parfeval** to offload a computationally demanding task to a parallel worker. Once the background task is completed, it updates the user interface without freezing the application. Below, we describe the code, plot (Fig. 9.2), and optimization techniques used:

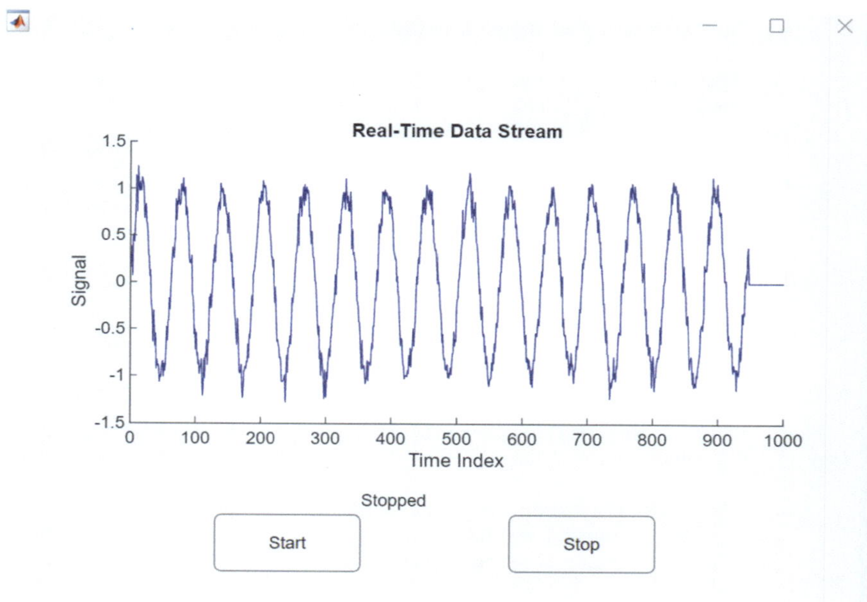

Fig. 9.2 Performance optimization example—real-time with background computation: Start and Stop options

```
classdef RealTimePlotApp_Background < matlab.apps.AppBase

    % App components
    properties (Access = public)
        UIFigure       matlab.ui.Figure
        UIAxes         matlab.ui.control.UIAxes
        StartButton    matlab.ui.control.Button
        StopButton     matlab.ui.control.Button
        StatusLabel    matlab.ui.control.Label
    end

    % Internal properties
    properties (Access = private)
        TimerObj      timer
        DataBuffer    double
        BufferSize    = 1000  % Preallocated buffer size
        Index         = 1
        PlotLine      matlab.graphics.chart.primitive.Line
        Pool          parallel.Pool
    end

    methods (Access = private)

        % Start data streaming
        function startStreaming(app)
            % Preallocate buffer
            app.DataBuffer = zeros(1, app.BufferSize);
            app.Index = 1;

            % Ensure parallel pool exists
            if isempty(gcp('nocreate'))
                app.Pool = parpool('threads'); % lightweight pool
            end

            % Create timer for real-time updates
            app.TimerObj = timer( ...
                'ExecutionMode','fixedRate', ...
```

```
                'Period',0.05, ...
                'TimerFcn',@(~,~)app.updatePlot);

            start(app.TimerObj);
            app.StatusLabel.Text = "Streaming...";
        end

        % Stop data streaming
        function stopStreaming(app)
            if ~isempty(app.TimerObj) && isvalid(app.TimerObj)
                stop(app.TimerObj);
                delete(app.TimerObj);
            end
            app.StatusLabel.Text = "Stopped";
        end

        % Update plot efficiently
        function updatePlot(app)
            % Simulate new data point
            newData = sin(0.1*app.Index) + 0.1*randn;

            % Store in buffer
            app.DataBuffer(app.Index) = newData;
            app.Index = app.Index + 1;

            % Reset index if buffer is full
            if app.Index > app.BufferSize
                app.Index = 1;

                % Launch background computation when buffer fills
                f = parfeval(@app.heavyComputation, 1, app.Data-
Buffer);

                afterEach(f, @(result)app.updateStatus(result), 0);
            end

            % Update plot line
            set(app.PlotLine, 'YData', app.DataBuffer);
        end

        % Heavy computation (runs in background)
        function result = heavyComputation(~, data)
            % Example: filter + RMS calculation
            d = designfilt('lowpassiir','FilterOrder',8, ...
                'HalfPowerFrequency',0.15,'DesignMethod','butter');
            filtered = filtfilt(d, data);

            % Simulate heavy work
            pause(0.5);

            % Calculate RMS of filtered signal
            result = sqrt(mean(filtered.^2));
```

```
        end

        % Update UI after background computation
        function updateStatus(app, result)
            app.StatusLabel.Text = sprintf("RMS: %.3f", result);
        end
    end

    % App initialization and UI components
    methods (Access = private)

        function createComponents(app)
            % Create UIFigure
            app.UIFigure = uifigure('Position',[100 100 600 400]);

            % Create UIAxes
            app.UIAxes = uiaxes(app.UIFigure, ...
                'Position',[50 100 500 250]);
            xlabel(app.UIAxes,'Time Index');
            ylabel(app.UIAxes,'Signal');
            title(app.UIAxes,'Real-Time Data Stream');
          % Create buttons
            app.StartButton = uibutton(app.UIFigure,'push', ...
                'Text','Start', ...
                'Position',[150 30 100 40], ...
                'ButtonPushedFcn', @(~,~)app.startStreaming);

            app.StopButton = uibutton(app.UIFigure,'push', ...
                'Text','Stop', ...
                'Position',[350 30 100 40], ...
                'ButtonPushedFcn', @(~,~)app.stopStreaming);

            % Status label
            app.StatusLabel = uilabel(app.UIFigure, ...
                'Position',[250 70 200 20], ...
                'Text','Idle');

            % Precreate plot line
      app.PlotLine = plot(app.UIAxes, nan(1,app.BufferSize), 'b-');
        end
    end

    methods (Access = public)
        % Constructor
        function app = RealTimePlotApp_Background
            createComponents(app);
            registerApp(app, app.UIFigure);
        end
    end
end
```

The following are the optimization techniques carried out by this method:

- **Preallocation** avoids memory growth problems.
- **PlotLine** is reused with updated YData for **efficient plotting**.
- **Parfeval** performs **background parallel computation** to prevent UI updates from being blocked by filtering and RMS.
- **Responsive UI**: When the computation is complete, update the status label.

Here, during the execution of the function, we observe the following output in the MATLAB workspace:

>> RealTimePlotApp_Background

>> RealTimePlotApp_Background

ans =

RealTimePlotApp_Background with properties:

UIFigure: [1 × 1 Figure]
UIAxes: [1 × 1 UIAxes]
StartButton: [1 × 1 Button]
StopButton: [1 × 1 Button]
StatusLabel: [1 × 1 Label]

Starting parallel pool (parpool) using the "threads" profile ...
Connected to a **parallel pool with 8 workers**.

The results demonstrate how MATLAB performance can be optimized using "parallel pool" applications, which are particularly useful for running MATLAB on systems with limited memory and resources.

9.3 Caching

Caching involves the storage of frequently accessed information or the results of computationally intensive operations in a temporary location, referred to herein as the "cache," so that when the data is needed, it may be quickly accessed from the cache, thus saving the need for repeated computations or database fetches, and hence shortening the computation time [4, 6].

One technique that is often used for "caching" is **memorization,** in which the results of function calls are stored. When the same inputs occur again, the cached result is returned instead of executing the function again [7, 8].

An example of such a function is the Fibonacci sequence. The naive recursive implementation of such a function could be highly inefficient because it repeatedly calculates the exact Fibonacci sequence:

```
function f = fibonacci_naive(n)
    if n <= 1
        f = n;
    else
        f = fibonacci_naive(n-1) + fibonacci_naive(n-2);
    end
end
```

The above can be improved using caching (memorization):

```
function f = fibonacci_cached(n)
    persistent memo; % Declare memo as a persistent variable
    if isempty(memo)
        memo = containers.Map('KeyType', 'double', 'ValueType',
'double');
    end

    if memo.isKey(n)
        f = memo(n); % Return cached value
    elseif n <= 1
        f = n;
        memo(n) = f; % Store base case
    else
        f = fibonacci_cached(n-1) + fibonacci_cached(n-2);
        memo(n) = f; % Store computed value
    end
end
```

In **fibonacci_cached**, there is a persistent **containers.Map** named memo where previous results of the Fibonacci numbers have been stored. As soon as the function call **fibonac-ci_cached(n)** is made, it is checked if the value of "n" is a key in the memo. If so, the function returns the value because the result has already been cached. Otherwise, the value is calculated, stored in the memo, and finally returned. This certainly helps increase the function's performance on inputs [7, 8].

9.4 Profiling

Profiling is a method for analyzing a program's performance to identify its performance bottleneck. MATLAB's built-in tool **Profiler** provides insight into which portions of your code consume most of the execution time and thus helps you focus optimization efforts appropriately [4, 6, 7].

Below, we will use the naive Fibonacci example to illustrate the Profiler.

1. **Run the Profiler**: You can start the Profiler from the MATLAB command window:

```
>> profile on
fibonacci_naive(20); % Or a larger number like 30 for more
noticeable results
profile viewer
```

Alternatively, you can go to the **Home** tab in the MATLAB IDE, click **Run and Time**, and then select **Run and Time** from the dropdown.

Report Analysis: Upon completion, the Profiler opens an HTML report in your browser. This report summarizes where time was spent, including

- **Functions**: A list of all functions called, ordered by the time spent in them (inclusive time) or the time spent directly in them (exclusive time).

Perform Call Details: Clicking a function name provides a detailed view, including lines of code that took the most time consumption, the number of times each executed, and details regarding the function calls from within that function.

Looking at the Profiler report (Fig. 9.3) for **fibonacci_naive(20)**, you can see that most of the execution time was spent inside the **fibonacci_naive** function itself. Zooming in, you can clearly see that the recursive calls, namely **(fibonacci_naive(n-1) + fibonacci_naive(n−2))**, were responsible for such heavy computation [7, 8]. This highlights that the redundant calculations are the primary bottleneck in some mathematics-based computations.

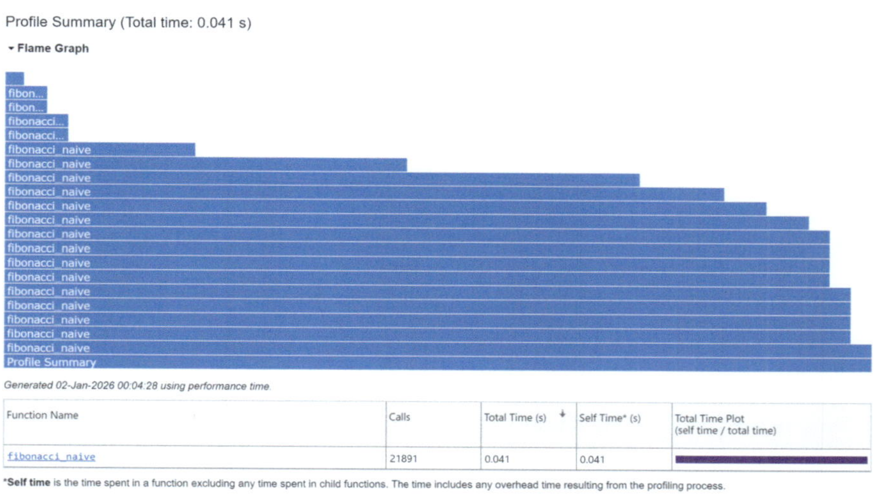

Fig. 9.3 MATLAB profiler example using "fibonacci_naive" function

Hence, the **Profiler** addresses questions such as

- Which functions are invoked most frequently?
- What lines of code are taking the longest to execute?
- Are there any unexpected or inefficient operations happening?

By identifying bottlenecks in your code, you can implement optimization techniques such as caching, *pre-allocation*, *vectorization*, and *algorithmic optimization* to enhance its performance. For example, from profiling **fibonacci_naive**, you would notice the necessity of using **fibonacci_cached** to improve its performance.

9.5 Conclusion

Performance optimization is a core activity for creating efficient and user-friendly MATLAB applications. Profiling, caching, and memory management-some strategies toward finding and resolving performance issues can bring significant improvements to both computational speed and user experience in MATLAB App. Not only do optimized MATLAB Apps run faster and process larger datasets more efficiently, but they are also much more professional and reliable. By finally adopting systematic optimization approaches, MATLAB-based tools will scale up, become easier to maintain, and support more advanced analytics and visualization tasks in contemporary engineering and research environments.

References

1. Akan A, Chaparro LF. Signals and systems using MATLAB®. Academic Press; 2024.
2. Attaway DC. Matlab: a practical introduction to programming and problem solving. 6th ed. Butterworth-Heinemann; 2022.
3. Darweesh M. Teaching MATLAB Programming by Utilizing Image Processing and Graphical User Interface Basics. Interactive Mobile Communication, Technologies and Learning. Springer; 2021. p. 824–32.
4. Hossain E. App Designer and Graphical User Interface in MATLAB. MATLAB and Simulink Crash Course for Engineers. Springer; 2022. p. 299–316.
5. Lent CS. Learning to program with MATLAB: Building GUI tools. John Wiley & Sons; 2022.
6. Mathworks: MATLAB App Designer: Create desktop and web apps in MATLAB. https://au.mathworks.com/products/matlab/app-designer.html (2025). Accessed September 2025.
7. Stahel A. Octave and MATLAB for engineering applications. Springer; 2022.
8. Youvan DC. Computational Sequences for Enhanced Efficiency: A Novel Approach to Data Handling, Security, and Performance Optimization. 2024.

Chapter 10
App Packaging and Distribution

10.1 Introduction

The interactive visualization tools of MATLAB, especially with App Designer, allow you to develop dynamic applications (Apps) with Graphical User Interfaces (GUIs). These Apps enable users to interact with data and see immediate visual feedback [1, 2]. After an application is created, MATLAB offers tools for managing updates and maintenance, sharing it with others, and packaging it into a single file for simple distribution [3, 4]. App Packaging is the process of combining all of your MATLAB applications made with App Designer into a **single installable package (.mlappinstall)**, along with the necessary files, data, and dependencies [5, 6]. This chapter explores App Packaging and distribution within the MATLAB environment.

10.2 Components of App Packages

A MATLAB App package typically contains the following components, which are outlined in Table 10.1.

Supplementary Information The online version contains supplementary material available at https://doi.org/10.1007/978-981-95-8797-1_10.

G. R. Naik, *MATLAB App Development*, https://doi.org/10.1007/978-981-95-8797-1_10

Table 10.1 MATLAB App components

Components	Description
.mlapp file	Main App designer file
.m files	Supporting functions/scripts
.mat/.csv files	Data or configuration files
.fig/.jpg/.png	GUI figures or icons
App metadata	App name, description, author, version, etc.

10.3 Creating App Packages

When you create an App package in MATLAB, you put your App and all required files and dependencies into a single **.mlappinstall** file, making it easy to share. This makes it easier to share your App because users don't have to worry about managing the MATLAB path or individual files.

10.3.1 How to Create an App Package

1. **Develop Your App in App Designer**: First, create your interactive visualization or application using MATLAB's App Designer. This involves dragging and dropping User Interface (UI) components (such as buttons, sliders, plots, and tables) onto a canvas and writing MATLAB code to define their behavior.
   - **Example**: Let us say you have created a simple "UI Alert" App where a user can click on the UI Alert push button to display an alert and can close the App with an exit button.
2. **Open the App in App Designer**: Ensure your **`.mlapp`** file is open in App Designer:

   app = Simple_UI_Alert.mlapp; % Example App

3. **Access the Sharing Options**: In the App Designer toolstrip, navigate to the **Designer** tab. On the far right, you will find a **Share** section (Fig. 10.1).
4. **Select "MATLAB App"**: Click on **Share** and then choose **MATLAB App**. This will open the "Configure MATLAB App for Sharing" dialog box [2].
5. **Configure App Details**:
   - **App Appearance**: Provide details such as the app name, author, version, summary, and description. You can also specify a splash screen image (Fig. 10.2).
   - **Files Required for App to Run**: MATLAB automatically detects most dependencies. You can review and manually add any additional files your app might need (e.g., data files, helper functions not directly called by the App).

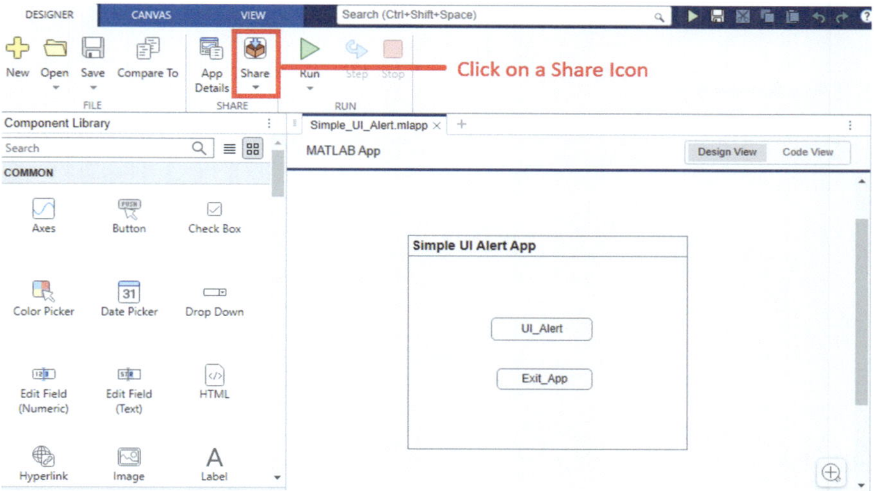

Fig. 10.1 MATLAB App sharing options

- **Manage App Sharing with MATLAB Project**: In most cases, packaging your App directly from App Designer is the easiest way to create an app artifact to share. However, certain packaging configuration options, such as specifying supported platforms or MATLAB releases and saving configurations for re-deployment, are available only when you use a project. The packaging tool will identify the MATLAB toolboxes required to run your app.

6. **Package the App**: Click the **Package** button. MATLAB will then compile and package your app into a. **mlappinstall** file (Fig. 10.3).
   - **Example Output**: After packaging, you will have a file like "**Simple_UI_Alert**" in the output folder. This file can then be shared with others [2].

10.3.2 Sharing Apps with Others

Once you have created an app package, there are several ways to share it with colleagues, collaborators, or the broader MATLAB community. The method you choose depends on whether your users have MATLAB installed and how widely you intend to distribute the app.

1. **Sharing the .mlappinstall File Directly**:
   - This is the simplest method for sharing with other MATLAB users.
   - **How it works**: You send the **.mlappinstall** file (e.g., via email, cloud storage, or a network drive).

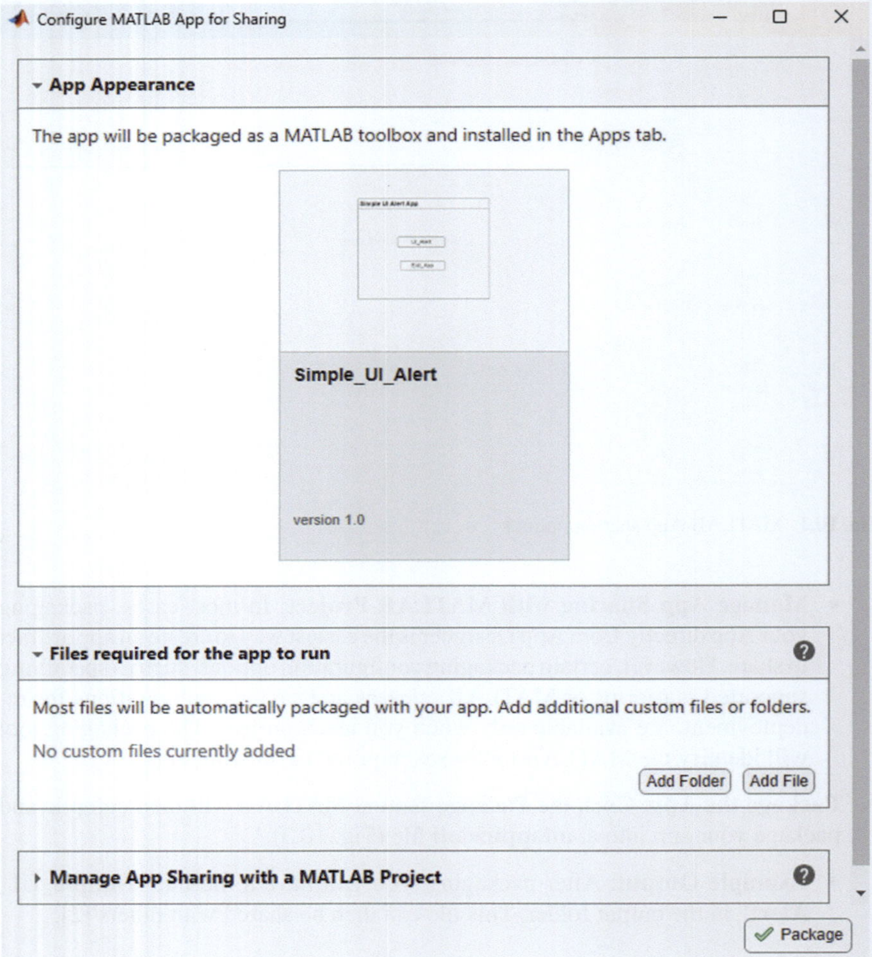

Fig. 10.2 Configuring MATLAB App package

- **User Installation**: The recipient **double-clicks** the **.mlappinstall** file in MATLAB. This triggers a one-click installation process, and the App will then appear in their **Apps tab** on the MATLAB toolstrip.
- **Requirement**: Users must have MATLAB installed and any required toolboxes that their App depends on.

2. **Sharing via MATLAB File Exchange**:

   - For wider distribution to the MATLAB community.

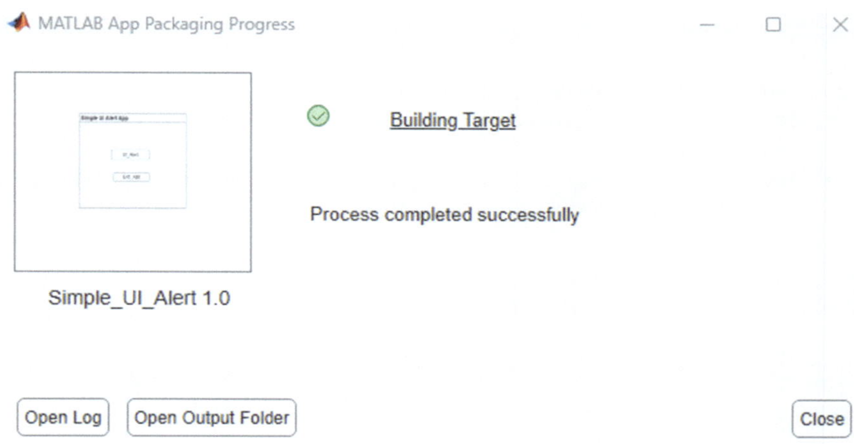

Fig. 10.3 Packaging MATLAB App

- **How it works**: You can upload your **.mlappinstall** file to the MATLAB File Exchange. This makes your App discoverable and downloadable by anyone with a MathWorks account.
- **User Installation**: Users can find and install your app directly from the MATLAB Add-Ons Explorer (**Home** tab > **Add-Ons** > **Get Add-Ons**).
- **Requirement**: Same as direct sharing, users need MATLAB and the relevant toolboxes.

3. **Creating Standalone Desktop Applications (requires MATLAB Compiler)**:

   - This allows you to share your App with users who **do not have MATLAB installed**.
   - **How it works**: Using **MATLAB Compiler**, you can compile your MATLAB App into a standalone executable (**.exe** for Windows, application bundle for macOS, etc.).
   - **User Requirement**: Users will need to install the **MATLAB Runtime**, a free set of shared libraries that allows compiled MATLAB applications to run.
   - **Benefits**: Ideal for deploying applications to a broader audience without MATLAB licenses.
   - **Process**: In App Designer, under the **Designer** tab > **Share**, select **Standalone Desktop App**. This will launch the Application Compiler.

4. **Creating Web Apps (requires MATLAB Compiler and MATLAB Web App Server)**:

   - Allows users to run your app directly in a web browser, eliminating the need for any local MATLAB installation.

- **How it works**: You compile your app into a web app using **MATLAB Compiler** and then deploy it to a **MATLAB Web App Server**. The server hosts the app, and users access it via a URL.
- **Benefits**: Cross-platform accessibility, no local installation for users, centralized deployment.
- **Process**: In App Designer, under the **Designer** tab > **Share**, select **Web App**. This will launch the Web App Compiler [2].

10.3.3 App Updates and Maintenance

It is essential to updating your MATLAB Apps to remove bugs, add new functionality, improve performance, and maintain compatibility.

Best Practices for App Updates:

1. **Version Control**:

   - **Use a version control system like Git**: This is very important for keeping track of modifications, collaborating with other people on your code, and reverting to a previous version if issues arise. You should place your **.mlapp** file (or your MATLAB App) and any other **.m** files that may be associated with your code under version control.
   - **Increment Version Numbers**: Always update the version number of the App whenever you have made significant changes or corrected some errors. This will make it easier for people to find the latest versions. When packaging the App, enter the version number in the "**Configure MATLAB App for Sharing**" dialog.

2. **User Feedback and Bug Reporting:**

   - **Create a feedback system**: Establish a feedback mechanism where you can take suggestions for improvements or bug fixes. This can be set up using an email address, an issue reporting system, or even a feedback mechanism if you build an application.
   - **Set Priorities for fixes**: Set priorities for fixes and address critical bugs to retain users' trust.

3. **Backward Compatibility**:

   - **Backward Compatibility as an Objective**: A key objective for implementing updates should be backward compatibility. There is an implication that if your application has progressed from an initial version to an advanced version, data or workflows from the initial version can still work with the advanced version. Changes that affect backward compatibility should be documented.

4. **Communication**:

   - **Provide updates to users**: Keep users informed about upgrades, modifications (also known as release notes), and how these changes are implemented. When submitting through the File Exchange, update your file with the modified **.mlappinstall** file included in the upload [2].

Releasing an Update:

1. **Modify your App**: Adjust your **.mlapp** file and other support functions that may be needed.
2. **Test Completely**: It is imperative to make sure that all the features of the application work well and that there are no new bugs in it. This is achieved by conducting a complete test of the application before its deployment (release).
3. **Repackage the App**: Make sure to increase the version number in the "**Configure MATLAB App for Sharing**" dialog by repeating the "**Creating App Packages**" steps.
4. **Distribute the updated package**:

   - **For .mlappinstall**: Give your users access to the updated **.mlappinstall** file. MATLAB will offer to update the current installed version when they double-click it.
   - **For standalone or Web Apps**: Use the MATLAB compiler to recompile the program and re-deploy it to the required environment (e.g., re-deploy the updated executable to user machines, re-deploy the updated web App to the web server) [2].

10.4 Conclusion

This chapter provides an in-depth discussion of App packaging in MATLAB, an efficient way to convert interactive applications developed in the App Designer environment into packages that others can easily share and install. App developers can easily share their Apps with other MATLAB users without disclosing the App's structure by packaging the App's related files (scripts, data, images, and metadata).

MATLAB helps create a **.mlappinstall** file that allows the software to be installed from the App tab. The approach ensures functional reliability and simple version management, while facilitating the distribution of software applications across academic, research, and industrial communities.

The App Packaging and Distribution in MATLAB fills the gap between App creation and usability, helping users share their computational tools in a professional, accessible way. Regardless of the intended purpose (e.g., classroom demonstrations, research tools, or industry applications), packaging enhances reusability, accessibility, and collaboration within the MATLAB ecosystem.

References

1. Hossain E. App Designer and Graphical User Interface in MATLAB. MATLAB and Simulink Crash Course for Engineers. Springer; 2022. p. 299–316.
2. Mathworks: MATLAB App Designer: Create desktop and web apps in MATLAB. https://au.mathworks.com/products/matlab/app-designer.html (2025). Accessed September 2025.
3. Akan A, Chaparro LF. Signals and systems using MATLAB®. Academic Press; 2024.
4. Attaway DC. Matlab: a practical introduction to programming and problem solving. 6th ed. Butterworth-Heinemann; 2022.
5. Darweesh M. Teaching MATLAB Programming by Utilizing Image Processing and Graphical User Interface Basics. Interactive Mobile Communication, Technologies and Learning. Springer; 2021. p. 824–32.
6. Thabit AA, Alzidi E, Mohammed MA, Palchyk P. Development of a System Based on a Graphical User Interface Using MATLAB Tools. 2024 35th Conference of Open Innovations Association (FRUCT): IEEE; 2024. p. 744–54.

Part V
Real-World Examples

Chapter 11
Interest Rate and Mortgage App

11.1 Introduction

This chapter demonstrates the fundamental principles of building basic MATLAB Apps using App Designer. Here we present two examples from scratch: (i) Interest Rate Calculator and (ii) Mortgage Calculator. In the subsequent chapter, we will explore more advanced concepts and techniques to create sophisticated and interactive applications.

11.2 App Designer from Scratch

Here, we present some basic App Designer concepts before designing App Designer from scratch. The whole process involves the following:

1. **Create a new app:** In App Designer, click on a Blank App.
2. **Add UI components:** Add UI components (Text boxes, Buttons, Lists, Selections, etc.).
3. **Write the App code**:
    - Create callbacks for each operation button.
    - In each callback function
    - Read the input values from the text boxes.
    - Perform the calculation.
    - Display the result in the output text box.

Supplementary Information The online version contains supplementary material available at https://doi.org/10.1007/978-981-95-8797-1_11.

G. R. Naik, *MATLAB App Development*, https://doi.org/10.1007/978-981-95-8797-1_11

4. **Run and test the App**: Run the App and test its functionality by entering numbers and performing calculations [1–3].

We now explain some concepts for creating a new App and its components. Let us start this by creating **a new App**:

Option 1:

1. Go to the **Home** tab.
2. Click **New**.
3. Select **App** (refer to Fig. 11.1).

Option 2: Another way to access the App Designer is to navigate to the Apps menu and click "Design App" (see Fig. 11.2).

Step 2: Once you click on **App**, a new pop-up window will open. This window presents a range of App templates and pre-built examples to help you get started.

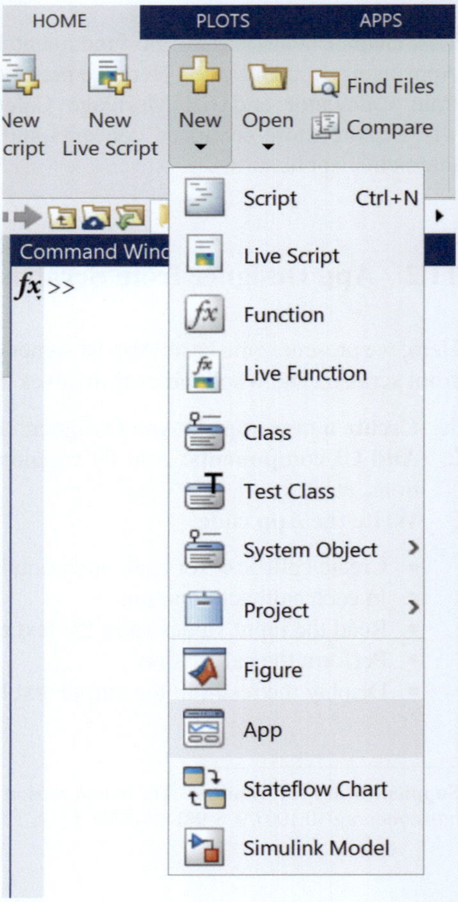

Fig. 11.1 Create a New App using the home tab

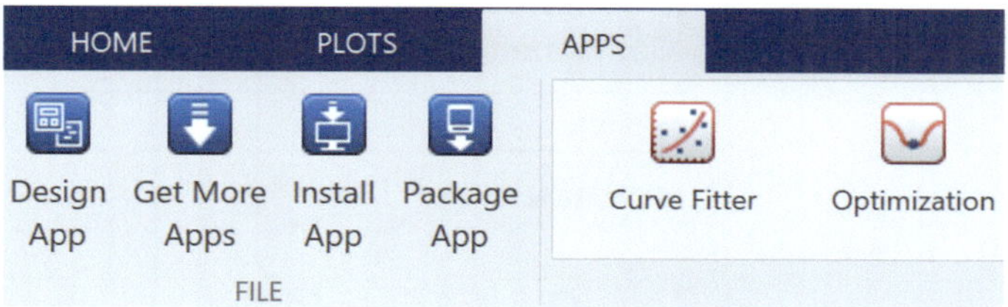

Fig. 11.2 Alternative way of designing MATLAB App

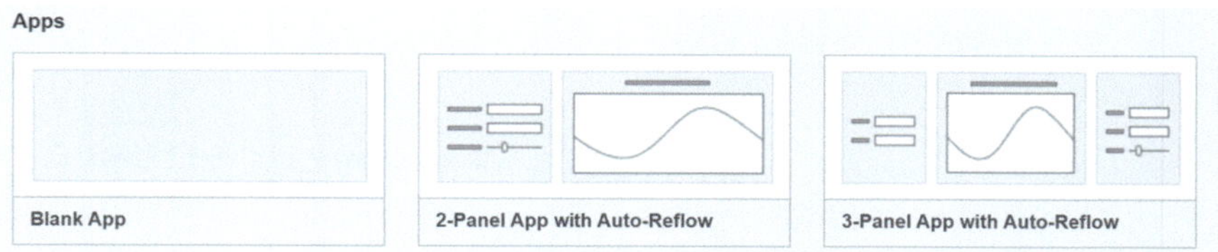

Fig. 11.3 MATLAB App template options

Step 3: Now, it's time to pick a template for your App. You can choose from three options (refer to Fig. 11.3):

- **Blank App**: Start from scratch and design your App exactly how you want it.
- **2-Panel App**: A simple layout with two sections, perfect for basic Apps.
- **3-Panel App**: A more complex layout with three sections, ideal for Apps with multiple features.

Step 4: Now, let's build your App's interface! The Component Library on the left side features numerous ready-made building blocks, including buttons, sliders, and graphs. Just drag and drop them onto your design area to create your App's look and feel (refer to Fig. 11.4).

Step 5: **MATLAB** consists of various components, such as

- Components
- Customizing Components

Components are pre-built shapes designed for specific tasks that can be imported into the design tab. In MATLAB, the *Component Library* is located on the left side of the window. To incorporate a component into your App's design, right-click the desired component in the Component Library, then drag it to the appropriate location on the design canvas.

The Component Browser, located to the right of the canvas design, allows you to customize various aspects of your components (refer to Fig. 11.5), including

- Font and color
- User interaction

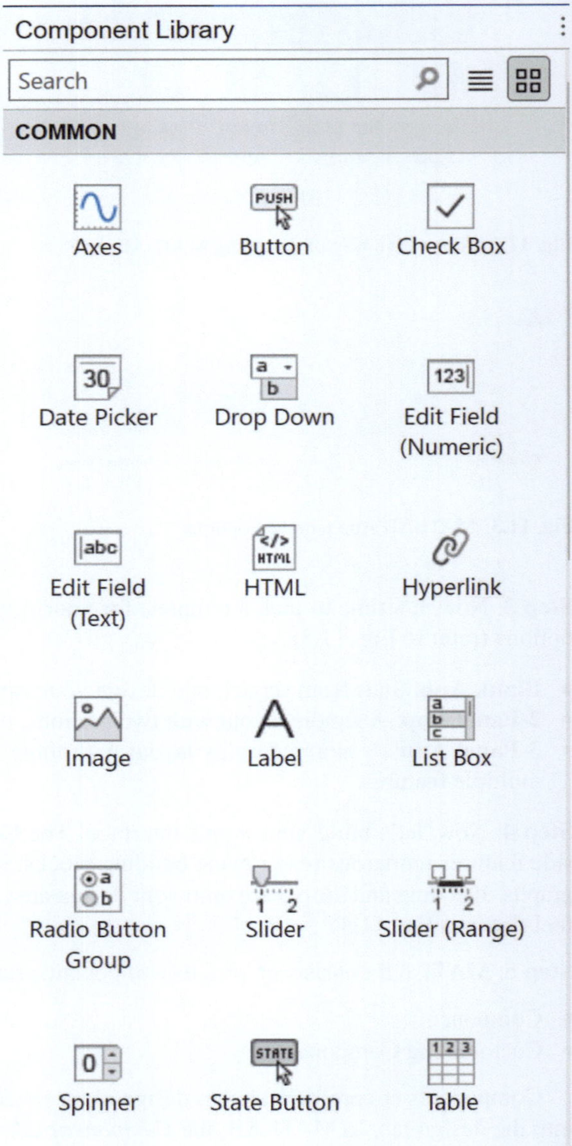

Fig. 11.4 MATLAB App component library options

- Position and size
- Callback functions
- Parent–child relationships
- Identifiers.

Now, let us focus on building a simple App by applying the above concepts.

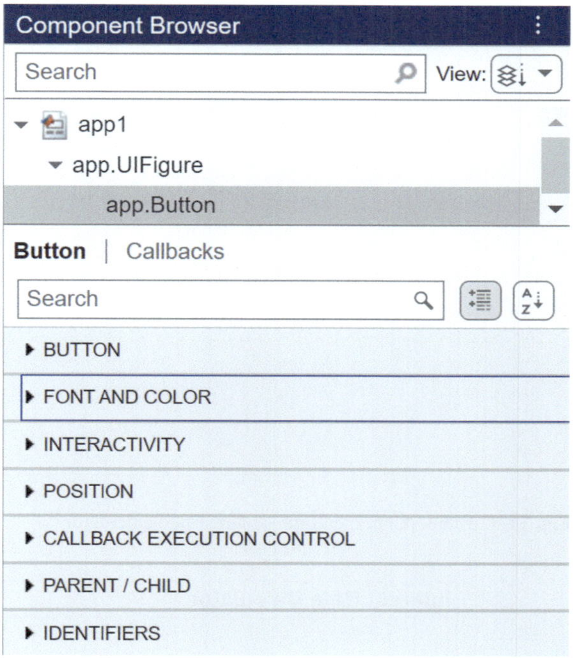

Fig. 11.5 App component browser customization options

11.2.1 Interest Rate Calculator

To gain practical experience with MATLAB App Designer, let us create a simple interest calculator App. Start by creating a new blank App.

Design the App's interface using the Design tab:

- Add a blank Panel from the component library and name it "Interest Rate Calculator" (refer to Fig. 11.6).
- Add five text fields: three for user input (Principal Amount, Interest Rate, and Time Period) and two for displaying the Simple Interest and Compound Interest (refer to Fig. 11.7).
- Add a button (push button) to initiate the calculation process (refer to Fig. 11.8).

The Code tab will be used to implement the calculation logic and update the output fields. To make the Calculate button perform the desired calculations, we will create a callback function. Right-click on the button, navigate to the **Callbacks** menu, and select **Add Callback**. This generates a function template that is executed whenever the button is pressed.

Now, add the following code to the "**CalculateInterestRateButtonPushed**" function:

Note: MATLAB App Designer automatically generates the required *function* and *end* code skeletons. Please paste the remaining code into the appropriate sections within these generated functions.

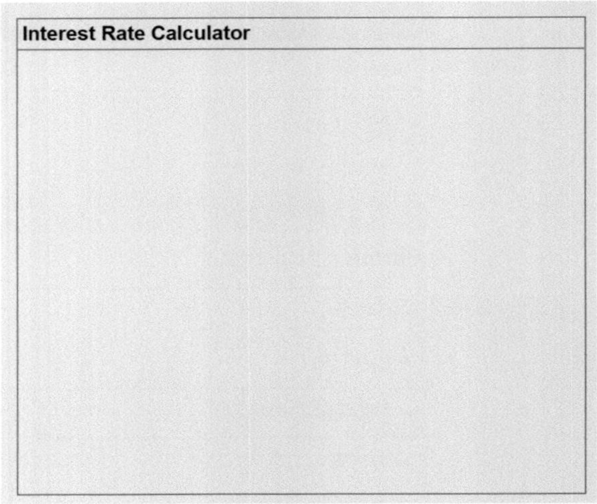

Fig. 11.6 Blank panel creation—Interest rate calculator

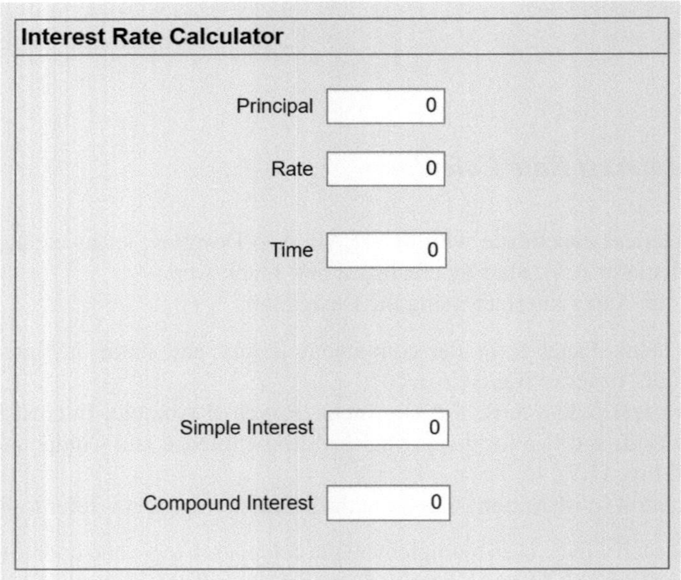

Fig. 11.7 Interest rate calculator button creation

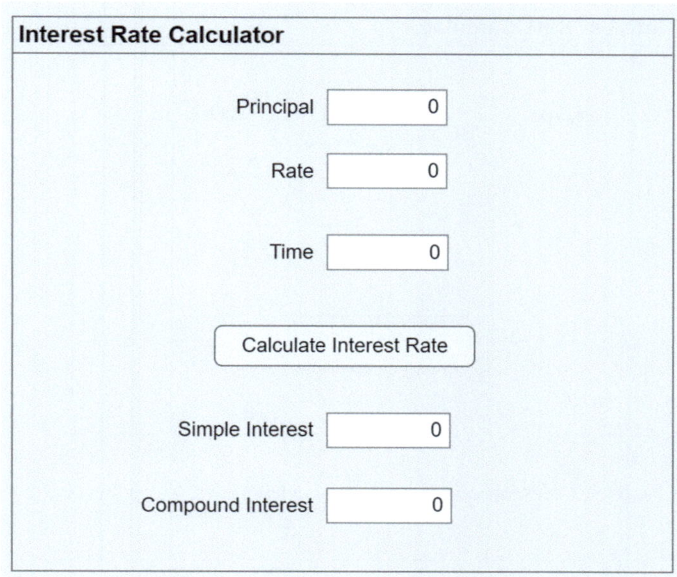

Fig. 11.8 Interest rate calculator—Push button creation

```
% Button pushed function: Calculate_Interest_RateBut-
ton
        function Calculate_Interest_RateButton-
Pushed(app, event)
            Amount = app.PrincipalEditField.Value;
              Rate = app.RateEditField.Value/100;
              Time = app.TimeEditField.Value;
              Simple_Interest = Amount*Rate*Time;
              app.SimpleInterestEditField.Value = Sim-
 ple_Interest;
              Compound_Amount = Amount *
 ((1+Rate).^Time);
              Compound_Interest = Compound_Amount -
 Amount;
              app.CompoundInterestEditField.Value =
 Compound_Interest;

          end
```

Upon inputting the data and executing the MATLAB App, we get the following output (refer to Fig. 11.9).

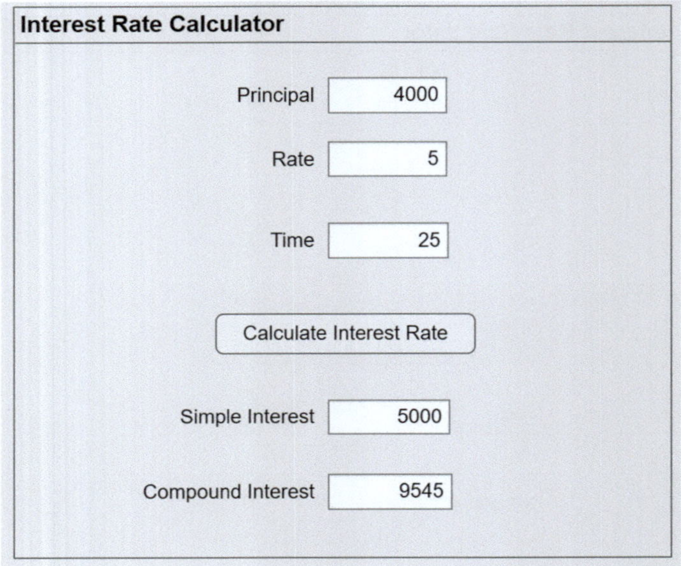

Fig. 11.9 Interest rate calculator—Working example

11.2.2 Modified App with Extra Features: Interest Rate Calculator

Now, let us modify the existing application by adding two extra features: (1) Clear the fields, and (2) Exit the application. To implement these, we will incorporate two push buttons: (1) Clear and (2) Exit (refer to Fig. 11.10). After adding these buttons, the application will have the following interface.

To make the Clear button perform the desired calculations, we will create a callback function. Right-click on the button, navigate to the **Callbacks** menu, and select **Add Callback**. This generates a function template that is executed whenever the button is pressed.

Add the following code to the "**ClearButtonPushed**" callback:

```
% Button pushed function: ClearButton
        function ClearButtonPushed(app, event)
            app.PrincipalEditField.Value = 0;
            app.TimeEditField.Value = 0;
            app.SimpleInterestEditField.Value = 0;
            app.CompoundInterestEditField.Value = 0;
            app.RateEditField.Value = 0;
        end
```

Similarly, add the following code to the "**ExitButtonPushed**" callback:

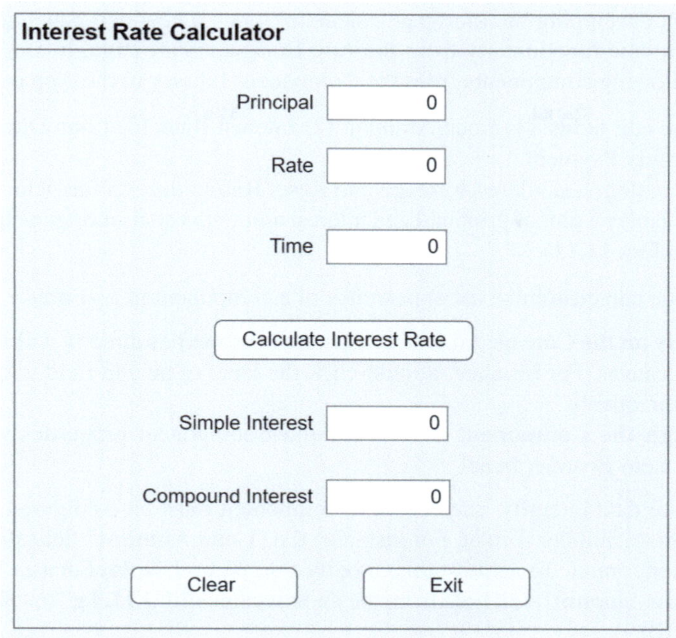

Fig. 11.10 Interest rate calculator—Modified App with extra features

```
% Button pushed function: ExitButton
       function ExitButtonPushed(app, event)
           app.delete % Shut down the entire app
       end
```

Once the results are executed, we can clear the edited field values and close the application (exit) at any time.

11.3 Mortgage Calculator

In App Designer, you can create an application that visualizes numeric data and dynamically updates based on user interactions [1, 4, 5]. Here, we design a simple Mortgage Calculator that accesses property values from user interface components and integrates data visualization within an application, which

- Collects user input.
- Calculates monthly payments based on the input.
- Generates a plot illustrating the distribution of principal and interest payments over time.
- Accessing property values from user interface components.

To begin developing the mortgage calculator, open a new two-panel application with auto-reflow functionality from the App Designer Start Page. In Design View, add the following components from the Component Library to the App canvas:

- Numeric edit fields: (1) Loan Amount, (2) Interest Rate, (3) Loan Duration, and (4) Monthly Payment.
- A push button: Calculate Mortgage and Reset Button to reset the field.
- Axes: Display a plot of principal and interest amounts versus mortgage instalment (refer to Fig. 11.11).

Next, you can customize the appearance of a component in two ways:

1. **Directly on the Canvas**: Edit the component's properties directly within the App design canvas. For instance, double-click the label of an edit field to modify its text interactively.
2. **Through the Component Browser**: Adjust component properties within the Component Browser panel.

To ensure data integrity, each editable component must be configured to accept the appropriate number format. For instance, the "**Loan Amount**" field should only allow numeric input. To achieve this, modify the "**ValueDisplayFormat**" property of the "**Loan Amount**" edit field from the default value of "**%11.4 g**" to "**%g**" (refer to Fig. 11.12).

To ensure the "Monthly Payment" is accurately represented, its format should accommodate both whole numbers and decimal values. To accomplish this, modify the "**ValueDisplayFormat**" property of the "**Monthly Payment**" edit field from the default setting of "**%11.4 g**" to "**%0.2f**." This change will limit the display of

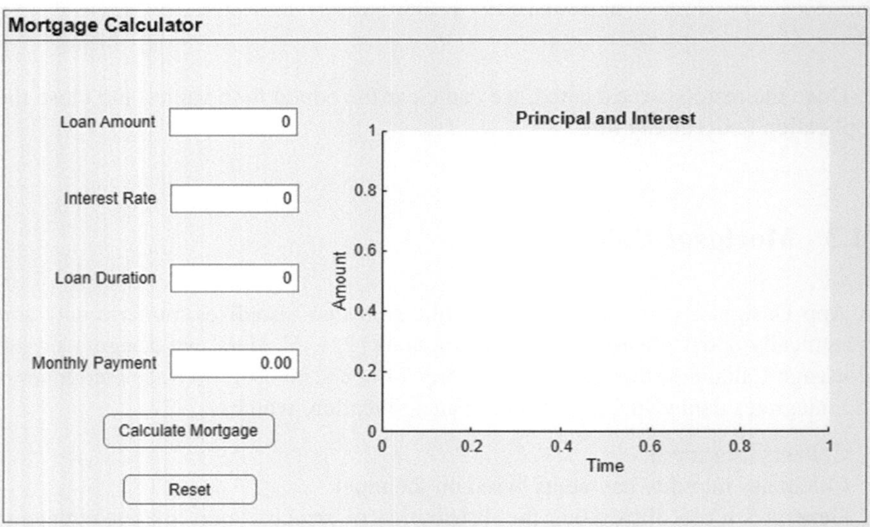

Fig. 11.11 Mortgage calculator App layout

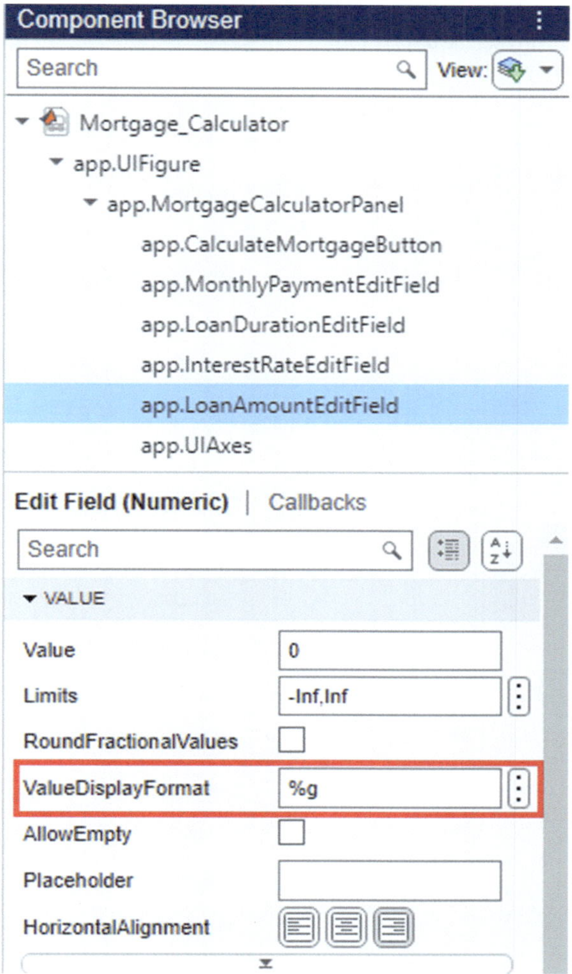

Fig. 11.12 An example of modifying the display format of the loan amount numeric edit field

decimal values to two places, providing a clear and concise representation of the monthly payment amount (refer to Fig. 11.13).

Similarly, edit other fields as required to handle integer and floating-point numbers up to the specified significant digits.

Define a Property for Internal Data

In *Code View*, under **Properties**, add:

```
properties (Access = private)
        PaymentSchedule table  % store payment break-
down
    end
```

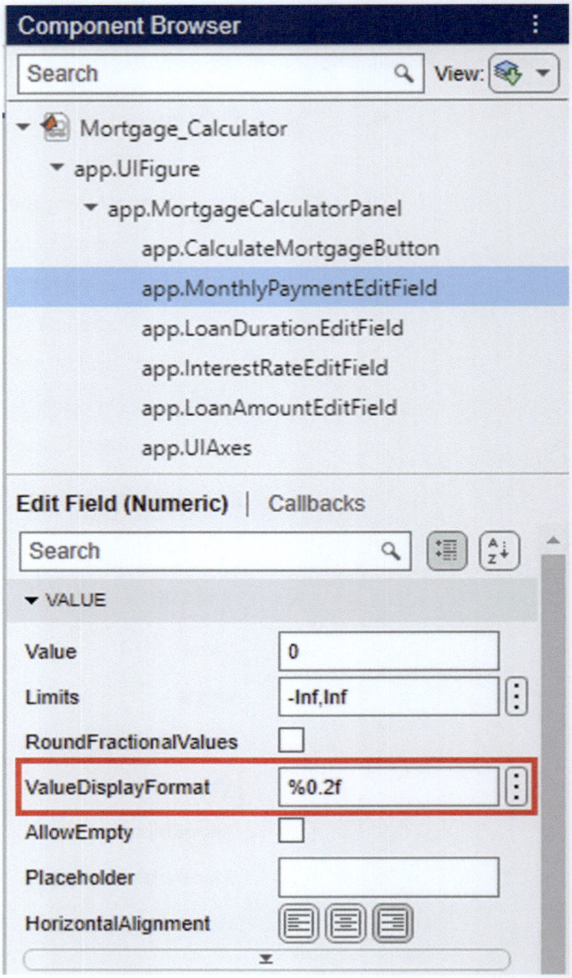

Fig. 11.13 An example of modifying the display format of the monthly payment edit field

Right-click on the "Calculate Mortgage" Button, and select the callback function. Enter the following code:

```
% Button pushed function: CalculateMortgageButton
function CalculateMortgageButtonPushed(app, event)
            % Button pushed function: ComputeButton

            % Get user inputs
            P = app.LoanAmountEditField.Value;
% principal
            annualRate = app.InterestRateEdit-
Field.Value / 100;  % annual %
            years = app.LoanDurationEditField.Value;
% loan term in years

           % Basic checks
           if P <= 0 || annualRate <= 0 || years <= 0
                uialert(app.UIFigure, 'Please enter
positive values for all fields.', 'Invalid Input');
                return;
            end

            % Monthly rate and number of payments
            r = annualRate / 12;
            n = years * 12;

            % Monthly payment formula (standard amor-
tization)
           M = P * (r * (1 + r)^n) / ((1 + r)^n - 1);

            % Display the monthly payment
            app.MonthlyPaymentEditField.Value = M;

            % Build amortization schedule
            balance = zeros(n, 1);
            interestPaid = zeros(n, 1);
            principalPaid = zeros(n, 1);
            remaining = P;

            for k = 1:n
                interestPaid(k) = remaining * r;
                principalPaid(k) = M - inter-
estPaid(k);
                remaining = remaining - principal-
Paid(k);
                balance(k) = max(remaining, 0);
            end

            months = (1:n)';
```

```
            app.PaymentSchedule = table(months, prin-
cipalPaid, interestPaid, balance, ...
                'VariableNames', {'Month', 'Princi-
pal', 'Interest', 'Balance'});

            % Plot results
            cla(app.UIAxes);
            hold(app.UIAxes, 'on');
            plot(app.UIAxes, months, app.Pay-
mentSchedule.Balance, 'LineWidth', 2);
            plot(app.UIAxes, months, app.Pay-
mentSchedule.Principal, '--', 'LineWidth', 1.2);
            plot(app.UIAxes, months, app.Pay-
mentSchedule.Interest, '--', 'LineWidth', 1.2);
            hold(app.UIAxes, 'off');

            grid(app.UIAxes, 'on');
            xlabel(app.UIAxes, 'Month');
            ylabel(app.UIAxes, 'Amount ($)');
            title(app.UIAxes, 'Mortgage Amortization
Schedule');
            legend(app.UIAxes, {'Remaining Balance',
'Principal Paid', 'Interest Paid'}, 'Location',
'best');

        end
```

Right-click on the "Reset" Button, and select the callback function. Enter the following code:

```
        % Button pushed function: ResetButton
        function ResetButtonPushed(app, event)
            app.LoanAmountEditField.Value = 0;
            app.InterestRateEditField.Value = 0;
            app.LoanDurationEditField.Value = 0;
            app.MonthlyPaymentEditField.Value = 0;
            cla(app.UIAxes);
        end
```

Now, save the App as "**Mortgage_Calculator.mlapp**" in the MATLAB directory. Upon running the App, enter the "loan amount," "Interest Rate," and "Loan Duration." The App displays the monthly amount due and provides a plot showing the remaining amount, principal, and monthly payment details (refer to Fig. 11.14). You can "Reset" and enter the new values to test the App.

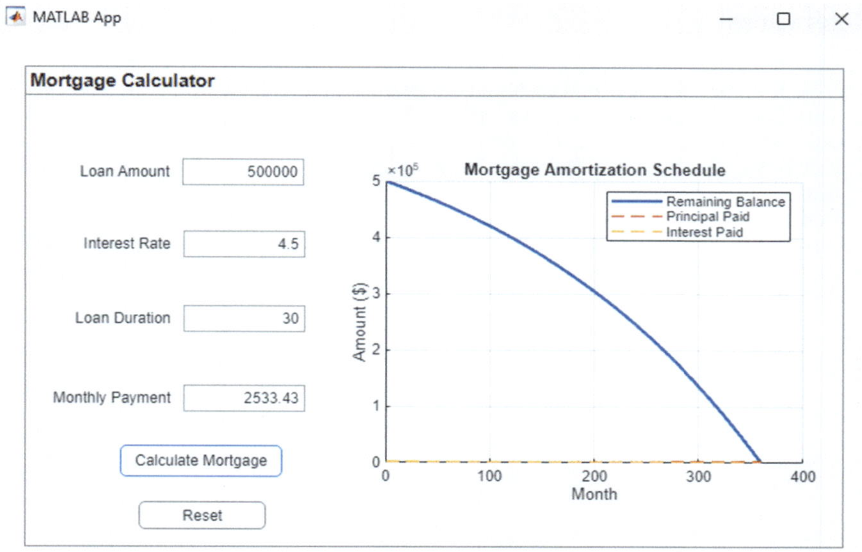

Fig. 11.14 Mortgage calculator App testing

11.4 Conclusion

In this chapter, we explored the foundational concepts of building basic MATLAB applications using App Designer. Through the development of two illustrative examples—(i) the Interest Rate Calculator and (ii) the Mortgage Calculator—we demonstrated how to design intuitive user interfaces, integrate computational logic, and create functional, interactive tools from scratch.

These examples provide a solid starting point for understanding the structure and workflow of MATLAB App development. The principles covered here lay the groundwork for more advanced topics presented in subsequent chapters, where we will expand on these fundamentals to build sophisticated, data-driven, and interactive applications.

References

1. Hossain E. App Designer and Graphical User Interface in MATLAB. MATLAB and Simulink Crash Course for Engineers. Springer; 2022. p. 299–316.
2. Lent CS. Learning to program with MATLAB: Building GUI tools. John Wiley & Sons; 2022.
3. Mathworks: MATLAB App Designer: Create desktop and web apps in MATLAB. https://au.mathworks.com/products/matlab/app-designer.html (2025). Accessed September 2025.
4. Akan A, Chaparro LF. Signals and systems using MATLAB®. Academic Press; 2024.
5. Darweesh M. Teaching MATLAB Programming by Utilizing Image Processing and Graphical User Interface Basics. Interactive Mobile Communication, Technologies and Learning. Springer; 2021. p. 824–32.

Chapter 12
Data Analysis App

12.1 Introduction

Data analysis is very important in making meaningful insights from data, facilitating scientific discovery, engineering, and informed decision-making [1–3]. MATLAB, with its powerful computational engine and interactive environment, is indeed the right software development platform for building applications that simplify data analysis. The Data Analysis App, designed in MATLAB, is a complete software package that combines data visualization, preprocessing, statistical modeling, and machine learning tools, all integrated into a user-friendly graphical user interface (GUI), making data analysis easy even without advanced programming knowledge [4–7].

This chapter introduces the design, development, and essential features of the Data Analysis App, which is based on MATLAB. The chapter shows how a simple data analysis can be expanded into a more complex data analysis App. The chapter provides information on how the App supports data loading, visualization, and manipulation. The App also enables real-time analysis, making it essential for numerous research applications such as biomedical signal processing, sensor data analysis, and system monitoring.

Supplementary Information The online version contains supplementary material available at https://doi.org/10.1007/978-981-95-8797-1_12.

G. R. Naik, *MATLAB App Development*, https://doi.org/10.1007/978-981-95-8797-1_12

12.2 Simple Data Analysis App

Here, we present a simple data analysis App capable of reading data from either a.txt or a.csv file. The App is also able to calculate key elements required for statistical analysis, which consist of (i) mean, (ii) standard deviation, (iii) max, and (iv) min:

Design Layout: Add a blank Panel from the component library and name it "Simple Data Analysis App" (Fig. 12.1).

Add a button group that lets the user select one of the four analysis options (mean, standard deviation, minimum, or maximum). Include one button to read the data (.csv or.txt). **browseFile** function: (i) Calls a file browser dialog box that enables the user to select the data file. (ii) Updates the **FileEdit** element with the user-supplied file path. Include one Edit element for the outcome and a UIAxes related to the graphical representation of the data (Fig. 12.2).

File Selection and Load Data: The "**Load_Data**" button uses the BrowseFile function to open a file dialog, allowing the user to select a data file. Here, we provide the option to load a.csv file and utilize the following code in the "**Load_Data**" callback function:

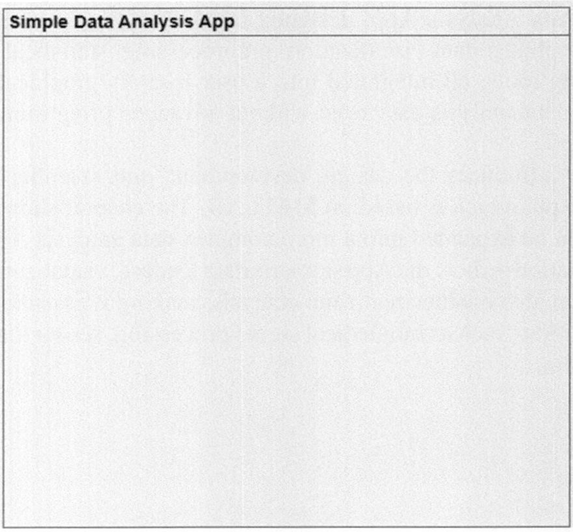

Fig. 12.1 Layout of Simple Data Analysis App

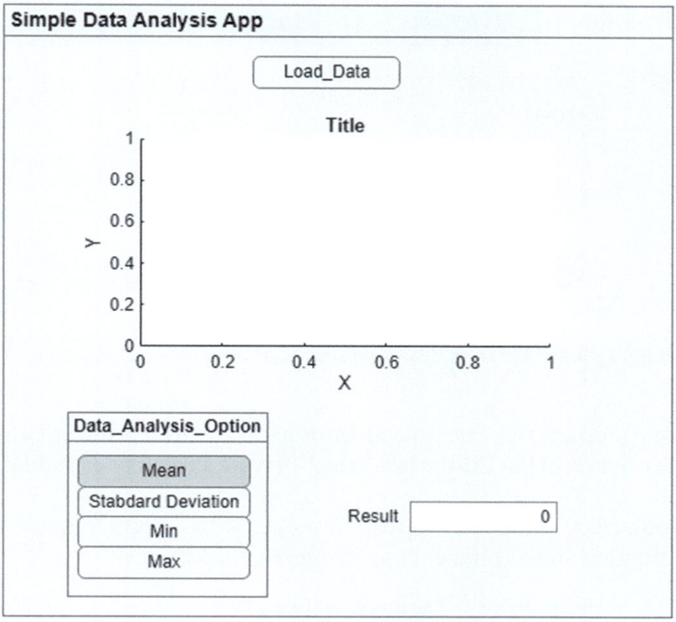

Fig. 12.2 Adding UIAxes and buttons to the Simple Data Analysis App

```
 % Button pushed function: Load_DataButton
        function Load_DataButtonPushed(app, event)
        file = uigetfile('*.csv'); % only let them
pick a .csv
            % Read the data into a table and save it
as a property
            % so it will be available to other
callbacks
            app.data = readtable(file);
        end
```

Here, to share data between various functions, we use the "**properties function**." Properties are the most effective way to share data within an app because they are accessible by all methods (functions and callbacks) within the app's class. This includes UI components, which you can access and update conveniently within your callback functions.

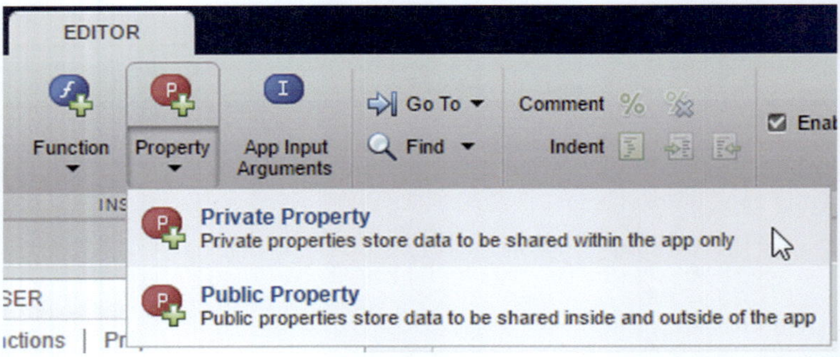

Fig. 12.3 Editor to choose the property of data (Public or Private)

To define a property for data, expand the drop-down menu from the bottom half of the **Property** button in the **Editor** tab. Select **Private Property** or **Public Property** (Fig. 12.3).

After you select an option to create a property, the App Designer will add a property definition and a comment to a **properties** block:

```
properties (Access = public)
        Property % Description
        data; %Data used for analysis
    end
```

The **properties block** is editable, so you can change the property name and edit the comment to describe the property, "`%Data used for analysis`". We can use "data" as app data for any function of this app. Once you define a property, you can access and set the property value anywhere in your app code using the syntax *app.PropertyName* (app.data).

Data Analysis Option: buttongroup allows users to select one analysis option (mean, standard deviation, minimum, maximum). We add the following code to the "Data_Analysis_Option" button group and UI plot:

```
% Selection changed function: Data_Analysis_Op-
tionButtonGroup
        function Data_Analysis_OptionButtonGroupSe-
lectionChanged(app, event)
            selectedButton = app.Data_Analysis_Op-
tionButtonGroup.SelectedObject;
```

```
            result = []; % Initialize result outside
the switch block

            switch selectedButton
                case app.MeanButton
                    result = mean(app.data{:,1}); %
app.data syntax is used to access the stored data
                case app.StandardDeviationButton
                    result = std(app.data{:,1});
                case app.MinButton
                    result = min(app.data{:,1});
                case app.MaxButton
                    result = max(app.data{:,1});
            end

            if isempty(result)
                % Handle empty result case (e.g.,
display a message)
                uialert(app, 'The analysis resulted
in an empty val-ue.', 'Information');
            else
                app.ResultEditField.Value = result;

                % Plot the data (assuming numerical
result)
                cla(app.UIAxes);
                bar(app.UIAxes, app.data{:,1});
                xlabel(app.UIAxes, 'Number of Data
Points');  % Adjust label accordingly
                ylabel(app.UIAxes, 'Value');
                title('Data used for Simple Statis-
tics Evaluation');
            end
        end
```

Run and use the App: Save the code as an `.mlapp` file (`Simple_Data_AnalysisApp.mlapp`) and Run the file in MATLAB.

1. Load a data file using the "Load_Data" button.
2. Select an analysis option from the button group.
3. The result will be displayed in the `ResultEdit`, and the data will be plotted in the `uiaxes` (Fig. 12.4).

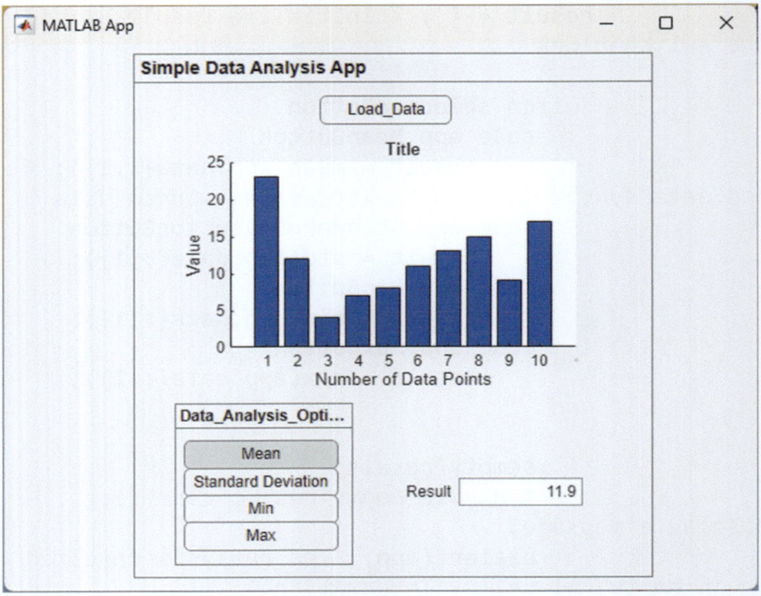

Fig. 12.4 Analysis results example of the Simple Data Analysis App

This simple data analysis app can only calculate basic statistical measures for a single column of.csv data. Next, we extend this App to perform advanced statistical computations.

12.3 Advanced Data Analysis APP

Here, we present an advanced Data analysis App that loads the data from a.csv file and computes both basic statistical elements, such as count, mean, standard deviation, max, and min and also advanced statistical parameters such as (i) range, (ii) skewness, (iii) kurtosis, (iv) Q1—25th (1st quartile), and (iv) Q4—75th (4th quartile).

The Advanced App can be created as follows:

- **Step 1**: As the first step, Open App Designer: in MATLAB, click **Home → New → App** or type **appdesigner** in the command window. Choose **Blank App** (or use a template). You will see *Design View* (left component palette, middle: canvas; right: Inspector).
- **Step 2: Drag components onto the canvas: Button** (from *Component Library → Button*), rename its Text (Inspector) to **LoadCSVButton**.
- **Dropdown**, rename label to **ColumnSelectDropdown**, and set Enable to off initially.

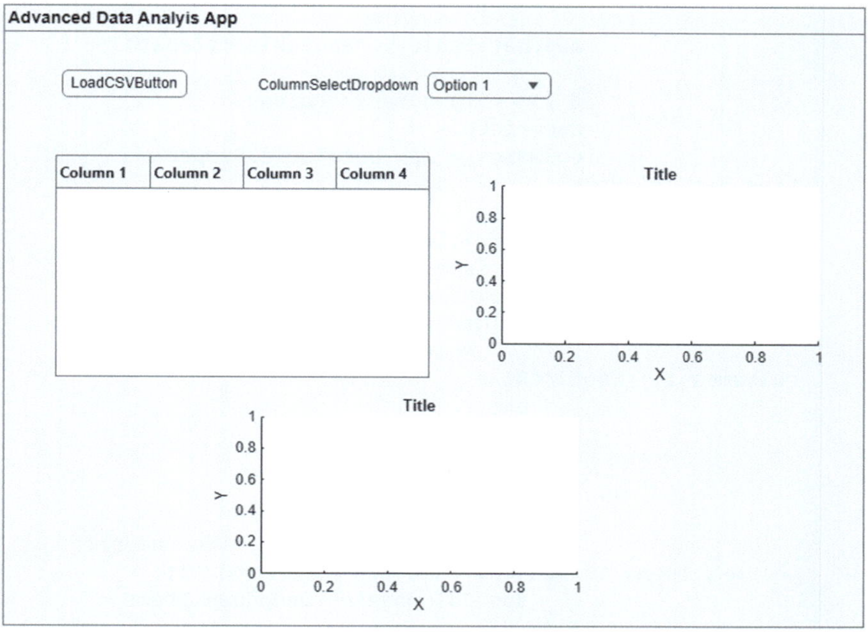

Fig. 12.5 UIAxes layout for displaying Histogram and Box plots

- Create a **UITable**, to show stats/results.
- Finally, create two **UIAxes**, one for displaying histogram and the other for displaying the boxplot (Fig. 12.5).
- **Step 3**: Select **LoadCSVButton**, right-click, and select Callbacks, and select Go to **LoadCSVButtonPushed callback.** App Designer will create the callback function and open *Code View*. Make sure to use `uigetfile` /`readtable` to load your CSV, populate `app.ColumnSelectDropdown.Items`: enable the dropdown and call its value-changed handler to update plots/tables. Add the following code:

```
% Button pushed function: LoadCSVButton
        function LoadCSVButtonPushed(app, event)
            [filename, pathname] = uigetfile('*.csv',
'Select a CSV file');
            if isequal(filename, 0)
                return; % User cancelled
            end

            filepath = fullfile(pathname, filename);
```

```
            try
                app.DataTable = readtable(filepath);

                % Identify numeric columns
                numericCols = {};
                colNames = app.DataTable.Proper-
ties.VariableNames;

                for i = 1:length(colNames)
                    if isnu-
meric(app.DataTable.(colNames{i})) && is-
real(app.DataTable.(colNames{i}))
                        numericCols{end+1} =
colNames{i}; %#ok<AGROW>
                    end
                end

                % Populate dropdown
                if isempty(numericCols)
                    uialert(app.UIFigure, 'No numeric
columns found in the CSV file.', 'Data Error');
                    app.ColumnSelectDropdown.Items =
{'No numeric data'};
                    app.ColumnSelectDropdown.Value =
'No numeric data';
                    app.ColumnSelectDropdown.Enable =
'off';
                else
                    app.ColumnSelectDropdown.Items =
numericCols;
                    app.ColumnSelectDropdown.Value =
numericCols{1};
                    app.ColumnSelectDropdown.Enable =
'on';

                    % Trigger analysis automatically
for first column
                    app.ColumnSelectDropdownValue-
Changed([]);
                end

            catch ME
                uialert(app.UIFigure, ['Error loading
file: ' ME.message], 'File Load Error');
            end
        end
```

- **Step 4**: Select **ColumnSelectDropdown**, right-click and select Callbacks, and select Go to **ColumnSelectDropdownValueChanged callback.** App Designer will create the callback function and open *Code View*. Add the following code to perform dropdown functions as well as all calculations for statistical measures that are described earlier:

```
% Value changed function: ColumnSelectDropdown
        function ColumnSelectDropdownValue-
Changed(app, event)
            columnName = app.ColumnSelect-
Dropdown.Value;

            if isempty(columnName) || is-
empty(app.DataTable) || strcmp(columnName, 'No nu-
meric data')
                app.ResultsTable.Data = {};
                cla(app.DistributionAxes);
                cla(app.QuartileAxes);
                return;
            end

            columnData = app.DataTable.(columnName);

            % Validate numeric data
            if ~isnumeric(columnData) || is-
empty(columnData)
                uialert(app.UIFigure, 'Selected col-
umn is not numeric or contains no valid data.', 'In-
valid Column');
                return;
            end

            % Calculate statistics
            stats = app.calculateAdvanced-
Stats(columnData);

            % Update Results Table
            app.ResultsTable.Data = formatStatsForTa-
ble(stats);
            app.ResultsTable.ColumnName = {'Metric',
'Value'};
            app.ResultsTable.ColumnWidth = {'auto',
'auto'};


            % --- Plot: Histogram ---
```

```
            cla(app.DistributionAxes);
            histogram(app.DistributionAxes,
columnData, 'FaceColor', [0.2 0.6 0.9], 'EdgeColor',
'k');
            title(app.DistributionAxes, ['Distribu-
tion of ' columnName], 'Interpreter', 'none');
            xlabel(app.DistributionAxes, columnName,
'Interpreter', 'none');
            ylabel(app.DistributionAxes, 'Frequen-
cy');
            grid(app.DistributionAxes, 'on');

            % --- Plot: Box Plot ---
            cla(app.QuartileAxes);
            boxplot(app.QuartileAxes, columnData,
'Notch', 'on', 'Whisker', 1.5, ...
                'Labels', {columnName}, 'Symbol',
'r.');
            title(app.QuartileAxes, ['Quartile and
Outlier Analysis: ' columnName], 'Interpreter',
'none');
            ylabel(app.QuartileAxes, columnName, 'In-
terpreter', 'none');
            grid(app.QuartileAxes, 'on');

            % Helper: format stats for display
            function tableData = formatStatsForTa-
ble(stats)
                metrics = fieldnames(stats);
                values = struct2cell(stats);
                formattedValues = cellfun(@(v)
num2str(v, '%.4f'), values, 'UniformOutput', false);
                tableData = [metrics, format-
tedValues];
            end
        end
    end

    % HELPER METHODS
    % ----------------------------------------------
----------------------
    methods (Access = private)
        function stats = calculateAdvancedStats(~,
data)
            data = data(~isnan(data));
```

```
            if isempty(data)
                stats = struct('Count', 0, 'Mean',
NaN, 'Median', NaN, 'StdDev', NaN, ...
                    'Min', NaN, 'Max', NaN, 'Range',
NaN, 'Skewness', NaN, ...
                    'Kurtosis', NaN, 'Q1_25th', NaN,
'Q3_75th', NaN);
                return;
            end

            try
                stats = struct( ...
                    'Count', numel(data), ...
                    'Mean', mean(data), ...
                    'Median', median(data), ...
                    'StdDev', std(data), ...
                    'Min', min(data), ...
                    'Max', max(data), ...
                    'Range', range(data), ...
                    'Skewness', skewness(data), ...
                    'Kurtosis', kurtosis(data), ...
                    'Q1_25th', prctile(data, 25), ...
                    'Q3_75th', prctile(data, 75));
            catch
                % In case Statistics Toolbox is miss-
ing
                stats = struct( ...
                    'Count', numel(data), ...
                    'Mean', mean(data), ...
                    'Median', median(data), ...
                    'StdDev', std(data), ...
                    'Min', min(data), ...
                    'Max', max(data), ...
                    'Range', range(data), ...
                    'Skewness', NaN, ...
                    'Kurtosis', NaN, ...
                    'Q1_25th', prctile(data, 25), ...
                    'Q3_75th', prctile(data, 75));
            end
        end
    end
```

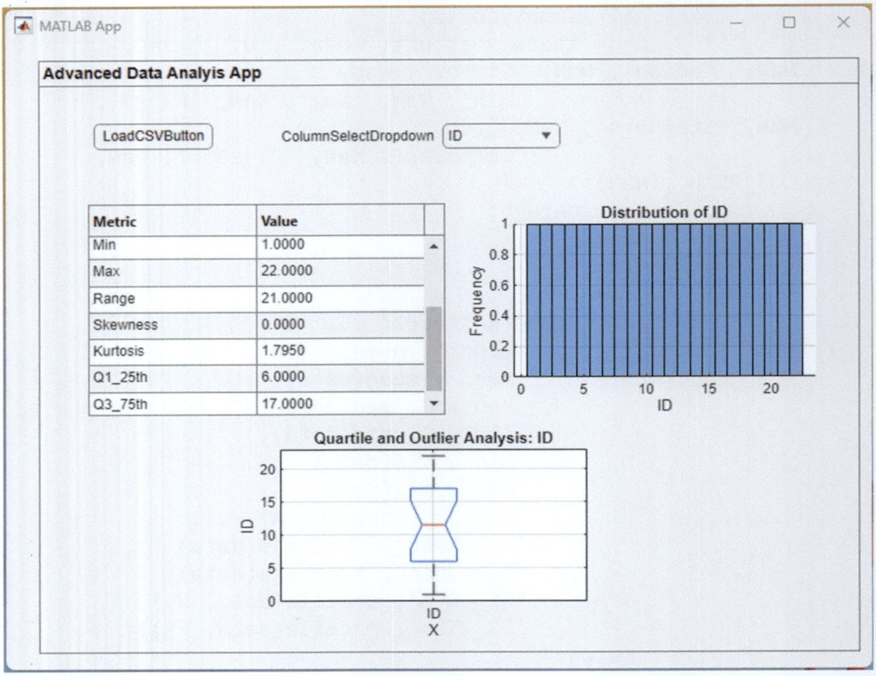

Fig. 12.6 Advanced data analysis App results—displaying results based on “ID” data

- **Step 5**: Add persistent data properties in the **Code Browser** under *Properties* and add the following code:

```
properties (Access = private)
    DataTable table  % Stores the loaded CSV data
end
```

- **Step 6**: Save the app as **Advanced_Data_Analysis.mlapp**. You can share the**.mlapp** file or package it (Apps → *Share App* → *Package* for installer/standalone with MATLAB Compiler). Press **Run** (green triangle) to test.

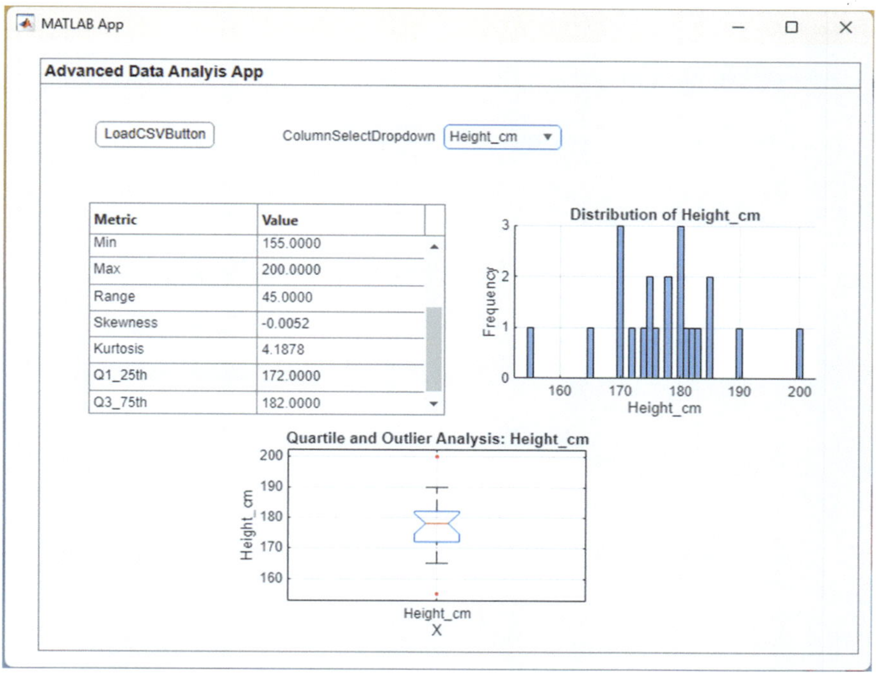

Fig. 12.7 Advanced data analysis App results—displaying results based on "Height_cm" data

For the test purpose, we create a sample data (.csv) file which contains the following: (i) ID, (ii) Height_cm, and (iii) Exam_scores. (Note: the sample file used for this plot is provided separately on the Springer website.) Please note that some statistical computations, such as skewness/kurtosis, require the **MATLAB Statistics Toolbox** and gracefully fall back if it is absent. Upon running the App, we can obtain the following results (Figs. 12.6, 12.2.7, and 12.8).

This App showcases how advanced data analysis can be performed in a MATLAB environment using custom-made Apps.

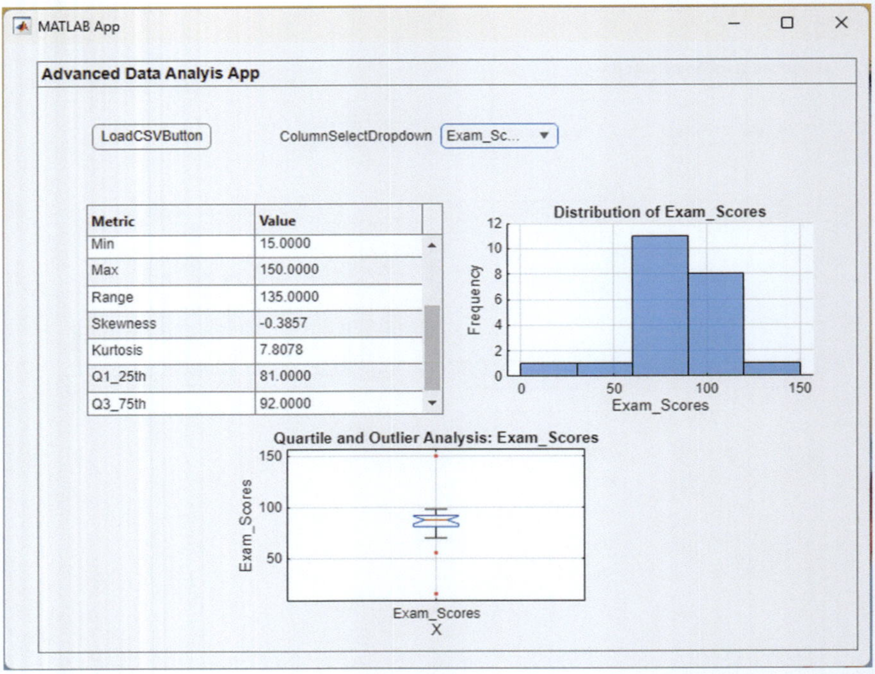

Fig. 12.8 Advanced data analysis App results—displaying results based on "Exam_Score" data

12.4 Conclusion

This chapter has demonstrated the creation and functionality of the Data Analysis Apps in MATLAB. Here, we illustrated MATLAB's interactive environment and its ability to convert data into information through a simple data analysis problem described in the early parts of this chapter, progressing to more complex examples presented in the later parts of the chapter.

This chapter has shown how such Apps facilitate real-time exploration and analysis of data, making them useful for various research and engineering applications, such as analyzing biomedical signals, interpreting sensor readings, and evaluating system performance. By presenting fundamental and advanced Data Analysis Apps, this chapter provides a comprehensive understanding of how to leverage MATLAB to develop analytical Apps.

References

1. Darweesh M. Teaching MATLAB Programming by Utilizing Image Processing and Graphical User Interface Basics. Interactive Mobile Communication, Technologies and Learning. Springer; 2021. p. 824–32.
2. Attaway DC. Matlab: a practical introduction to programming and problem solving. 6th ed. Butterworth-Heinemann; 2022.
3. Akan A, Chaparro LF. Signals and systems using MATLAB®. Academic Press; 2024.
4. Valentine DT, Hahn BH. Essential MATLAB for engineers and scientists. Academic Press; 2022.
5. Hossain E. App Designer and Graphical User Interface in MATLAB. MATLAB and Simulink Crash Course for Engineers. Springer; 2022. p. 299–316.
6. Lent CS. Learning to program with MATLAB: Building GUI tools. John Wiley & Sons; 2022.
7. Mathworks: MATLAB App Designer: Create desktop and web apps in MATLAB. https://au.mathworks.com/products/matlab/app-designer.html (2025). Accessed September 2025.

MIX
Papier aus verantwortungsvollen Quellen
Paper from responsible sources
FSC® C105338
FSC
www.fsc.org

If you have any concerns about our products,
you can contact us on
ProductSafety@springernature.com

In case Publisher is established outside the EU,
the EU authorized representative is:
Springer Nature Customer Service Center GmbH
Europaplatz 3, 69115 Heidelberg, Germany

Printed by Libri Plureos GmbH
in Hamburg, Germany